GEOMETRICAL COMBINATORIAL TOPOLOGY
Volume II

LESLIE C. GLASER
The University of Utah

VAN NOSTRAND REINHOLD COMPANY
LONDON
NEW YORK CINCINNATI TORONTO MELBOURNE

VAN NOSTRAND REINHOLD COMPANY
Windsor House, 46 Victoria Street, London, S.W.1

INTERNATIONAL OFFICES
New York Cincinnati Toronto Melbourne

Library of Congress Catalog Card No. 77–11338

ISBN 0 442 78283 7

First published 1972

Printed in Great Britain by
Butler & Tanner Ltd
Frome and London

Preface

The following lecture notes are a continuation of those given in Volume I and are the most interesting part of this work. Here, I assume that the reader is familiar with the ideas and techniques of Volume I, where I provide a complete and self-contained exposition of the foundations of geometrical combinatorial (piecewise linear) topology starting with the rudiments of the theory and then continuing to a number of important basic techniques that represent the transition period to modern geometrical combinatorial topology.

In Volume II, I give numerous applications to the theory developed in Volume I and present some of the earlier important results which illustrate additional basic techniques. The central theme is to provide the novice with as much of the flavor of the area as possible while using only a minimal amount of machinery. In particular, the emphasis is always placed on the geometry involved. As in the first volume, numerous exercises have been included so as to help the reader develop 'feeling' and facility in the subject. Thus the reader is only required to have some knowledge of point set topology and a familiarity with some of the more standard theorems in algebraic topology. Also, I have attempted to add a number of informal comments during various stages of many of the more complicated proofs or discussions so as to help the reader better understand what is involved or to remind him why a certain fact is true.

In Chapter I, I present a number of applications of Stallings' Engulfing Lemma. It includes sections covering the uniqueness of the PL structure of Euclidean space as developed by Stallings, the result that stable homeomorphisms can be approximated by piecewise linear ones as developed by Connell, and a discussion of Whitehead torsion and the Hauptvermutung as developed by Milnor.

In Chapter II, I consider numerous results related to embeddings of complexes in Euclidean space. Sections are included giving Zeeman's work on unknotting combinatorial balls, the work of a number of other geometrical topologists, including myself, involving pseudo-cells and contractible open manifolds, the basic work of Penrose, Whitehead and Zeeman on embedding manifolds in Euclidean space, and some of the work of Homma and Gluck related to the taming of complexes in Euclidean space.

Finally, for the convenience of the reader, I have included an appendix on locally flat embeddings of topological manifolds due to Morton Brown, an appendix giving some remarks on knot theory as developed by Fox, a basic list of references and select bibliography, and a complete index for both volumes.

L. C. GLASER

CONTENTS

CHAPTER I

Applications of the Engulfing Lemma

§A. The Piecewise Linear Structure of Euclidean Space

The ideas in this section are based on the paper by John Stallings, 'The Piecewise-Linear Structure of Euclidean Space', *Proceedings of the Cambridge Philosophical Society*, Vol. 58 (1962), pp. 481–488. The main results are as follows:

THEOREM I.1. *Let M^n be a contractible combinatorial n-manifold without boundary ($n \geqslant 5$) which is 1-connected at infinity, then M^n is pwl homeomorphic to Euclidean space E^n.*

COROLLARY I.2. *E^n has a unique pwl structure for $n \geqslant 5$.*

COROLLARY I.3: *Let X and Y be contractible open (non-compact without boundary) combinatorial manifolds such that* $\dim X \geqslant 1$, $\dim Y \geqslant 1$ *and* $\dim X + \dim Y \geqslant 5$, *then $X \times Y$ is pwl homeomorphic to E^n, where* $n = \dim X + \dim Y$.

COROLLARY I.4. *If M^n is a closed combinatorial homotopy n-sphere, $n \geqslant 5$, then M^n-{point} is pwl homeomorphic to E^n and hence M^n is topologically homeomorphic to S^n.*

First we will give some definitions and obtain some needed propositions. A space X is said to have one end, if X is not compact and if for every compact $C \subset X$, there exists a compact D

such that $C \subset D \subset X$ and $X - D$ is 0-connected (path connected). A space X is said to be 1-connected at ∞ if X is not compact and if for every compact $C \subset X$, there exists a compact D such that $C \subset D \subset X$ and $X - D$ is 1-connected (simply connected, or path connected with $\pi_1(X - D) = 0$). A space X or a pair (X, Y) is said to be k-connected if $\pi_i(X) = 0$ or $\pi_i(X, Y) = 0$ for all $i \leqslant k$.

We will assume Van Kampen's Theorem: Let A, B, $A \cap B$ be pathwise connected spaces, $x_0 \in A \cap B$ and suppose for every $x \in A \cup B$, that $x \in \text{int } A$ or $x \in \text{int } B$ with respect to $A \cup B$. Then $\pi_1(A \cup B, x_0)$ is the reduced free product of the following system

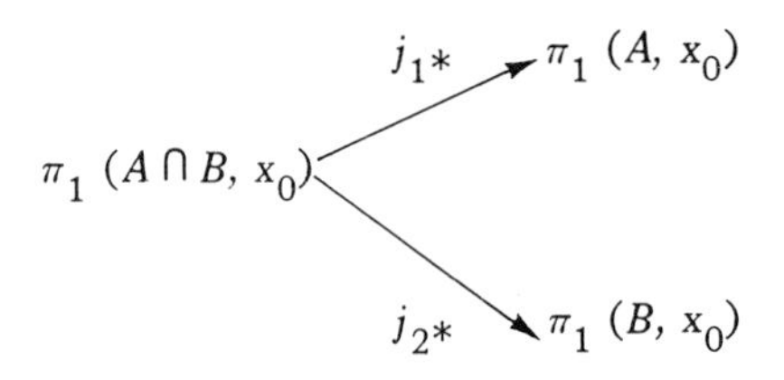

i.e., $\pi_1(A \cup B, x_0)$ is obtained from the free product of $\pi_1(A, x_0)$ and $\pi_1(B, x_0)$ by identifying the images in each of the elements of $\pi_1(A \cap B, x_0)$.

That is, $\pi_1(A \cup B, x_0) = \dfrac{\pi_1(A, x_0) * \pi_1(B, x_0)}{N}$ where N is the smallest normal subgroup generated by elements of the form $\{j_{1*}([f])\, j_{2*}([f^{-1}]) : [f] \in \pi_1(A \cap B, x_0)\}$.

PROPOSITION I.5. *If A and B are 1-connected spaces, each of which has one end, then $A \times B$ is 1-connected at infinity.*

Proof: Let C be a compact subset of $A \times B$. Since A and B each has one end, there exists D, E compact such that $D \subset A$, $E \subset B$, $C \subset D \times E$ and $A - D$ and $B - E$ are 0-connected.

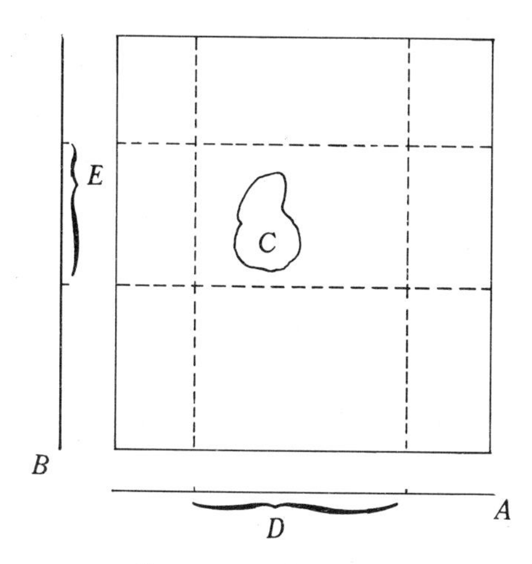

Now $(A\times B)-(D\times E) = [A\times(B-E)]\cup[(A-D)\times B]$ which we will write as $X\cup Y$. Then $[A\times(B-E)]\cap[(A-D)\times B] = (A-D)\times(B-E) = X\cap Y$. Each of the above 3-spaces X, Y, and $X\cap Y$ are open, 0-connected subsets of $(A\times B)-(D\times E)$.

Let j_1, j_2 be the inclusion of $X\cap Y$ into X and Y respectively and $p_0 = (a_0, b_0)\in X\cap Y$, where $a_0\in A-D$ and $b_0\in B-E$. Then we have $j_1\colon X\cap Y \longrightarrow X$, $j_2\colon X\cap Y\longrightarrow Y$ and the following induced diagram:

$$\pi_1((A-D)\times(B-E), p_0) \xrightarrow{j_{1*}} \pi_1(A\times(B-E), p_0) \underset{\cong}{\xrightarrow{\varphi}} \pi_1(A, a_0)\times\pi_1(B-E, b_0)$$

$$\pi_1((A-D)\times(B-E), p_0) \xrightarrow{j_{2*}} \pi_1((A-D)\times B, p_0) \underset{\cong}{\xrightarrow{\psi}} \pi_1(A-D, a_0)\times\pi_1(B, b_0).$$

Consider $[f]\in\pi_1(X)$ where $f\colon[0,1]\longrightarrow A\times(B-E)$ and $f(0) = f(1) = p_0$. Let p_1, p_2 be the projections of $A\times B$ onto A and B respectively. Now any $[f]\in\pi_1(X)$ is carried by φ to $[p_1 f, p_2 f] =$

$[a_0, p_2 f]$, since A is 1-connected and $[g] \in \pi_1(Y)$ is carried by ψ to $[p_1 g, p_2 g] = [p_1 g, b_0]$, since B is 1-connected. By Van Kampen's theorem $\pi_1(X \cup Y, p_0) \cong \dfrac{\pi_1(X, a_0) * \pi_1(Y, b_0)}{N}$ where N is the smallest normal subgroup of $\pi_1(X, a_0) * \pi_1(Y, b_0)$ generated by elements of the form $j_{1*}([h])\, j_{2*}([h]^{-1})$, $[h] \in \pi_1(X \cap Y, p_0)$.

Now consider any 'word' $W = [f_1][f_2]\ldots[f_n]$ in the free product. If $[f_i] \in \pi_1(X)$, then $[f_i] = [a_0, p_2 f_i]$ and if $[f_j] \in \pi_1(Y)$, then $[f_j] = [p_1 f_j, b_0]$. For $[f_i] \in \pi_1(X)$ consider $h = (a_0, p_2 f_i)$: $I \longrightarrow A - D \times B - E$; then $[h] \in \pi_1(X \cap Y, p_0)$ and $j_{1*}([h])\, j_{2*}([h]^{-1}) = [a_0, p_2 f_i][a_0, p_2 f_i^{-1}] = [a_0, p_2 f_i][a_0, b_0] = [a_0, p_2 f_i] = [f_i]$. In a similar fashion, if $[f_j] \in \pi_1(Y)$, letting $h' = (p_1 f_j^{-1}, b_0)$, we get $j_{1*}([h'])\, j_{2*}([h']^{-1}) = [f_j]$. Thus each $[f_i] \in N$ and by Van Kampen's Theorem it follows that $\pi_1(X \cup Y, p_0) = 0$. (Having given a rigorous use of Van Kampen's Theorem once, we will therefore omit future verifications.)

PROPOSITION I.6. *If A is a 1-connected space with 1 end and E^1 is the real line then $A \times E^1$ is 1-connected at ∞.*

Proof. Let C be a compact subset of $A \times E^1$, then there exists a compact $D \subset A$ such that $A - D$ is 0-connected and there exists a positive number $r \in E^1$, so that $C \subset D \times [-r, r]$. The claim is that $A \times E^1 - D \times [-r, r]$ is 1-connected.

Let $U = A \times (-\infty, r) - D \times [-r, r]$, $V = A \times (-r, \infty) - D \times [-r, r]$, and then $U \cap V = A - D \times (-r, r)$. Now $U = A \times (-\infty, -r) \cup (A - D) \times (-\infty, r)$, so by applying Van Kampen's Theorem to U, we get that $\pi_1(U) = 0$.

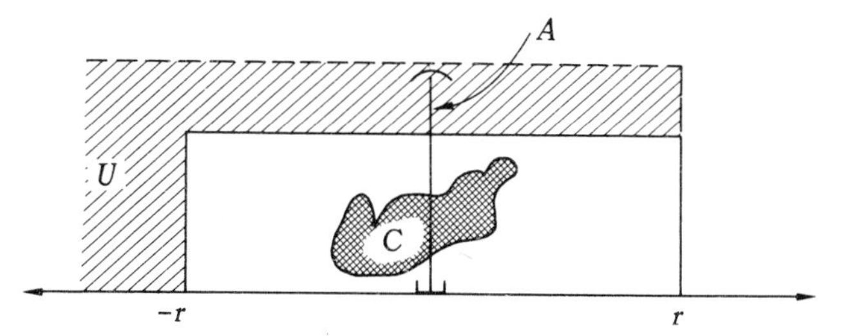

Similarly, $\pi_1(V) = 0$. Finally, applying Van Kampen's Theorem to $U \cup V$ we get that $\pi_1(U \cup V) = 0$.

PROPOSITION I.7. *A contractible manifold M^n without boundary of dimensions $n \geqslant 2$ has one end.*

Assuming this proposition for the moment we immediately obtain:

PROPOSITION I.8. *If $M^n = X \times Y$, where X and Y are manifolds without boundary, neither of which is a point, and if M is contractible and $n \geqslant 3$, then M is 1-connected at infinity.*

Proof. If each of X and Y are of dim $\geqslant 2$, Proposition I.7 implies that each of X and Y has 1 end and Proposition I.5 implies that $X \times Y$ is 1-connected at ∞. If either X or Y has dimension 1, then it is homeomorphic to E^1 and the other has dim $\geqslant 2$, then Proposition I.7 and Proposition I.6 imply $X \times Y$ is 1-connected at ∞ also.

Now we return to a proof of Proposition I.7. First we will need some more definitions and a lemma. Let K be a locally finite simplicial complex. If we remove from K a finite subcomplex L, we are left with $n(L)$ infinite components of $|K| - |L|$, where $n(L)$ is an integer. *Definition:* The number of ends of K, $e(K)$, is the l.u.b. of the integers $n(L)$. *Remark:* When removing such an L from K, we shall always enlarge L so that all components of $|K|-|L|$ are infinite (i.e., include all compact components of $|K|-|\text{original } L|$ with the new L).

Let $C^*(K)$ be the unrestricted cochains on K over $\mathbf{Z}_2$ where $C^n(K) = \operatorname{Hom}(C_n(K), \mathbf{Z}_2)$ and $C_n(K)$ is generated by n-simplexes of K. Let $C^*_f(K)$ be the finite cochains, i.e. all $c^n \in C^n(K)$ such that $c^n = 0$ except for a finite number of n-simplexes of K. Then we have the exact sequence:

$$0 \longrightarrow C_f^*(K) \longrightarrow C^*(K) \longrightarrow \frac{C^*(K)}{C_f^*(K)} = C_e^*(K) \longrightarrow 0.$$

In the usual way we get the exact sequence:

$$0 \longrightarrow H_f^0(K) \longrightarrow H^0(K) \longrightarrow H_e^0(K) \longrightarrow H_f^1(K) \longrightarrow H^1(K) \longrightarrow \cdots$$

LEMMA I.9. *The dimension of the vector space $H_e^0(K)$ is equal to the number of ends of K.*

Proof. Any element of $H_e^0(K)$ is a 0-cocycle in $C_e^0(K)$ and is represented by a 0-cochain $c \in C^0(K)$ whose coboundary is a finite 1-cochain, i.e. δc is zero outside some finite complex L. Hence c is constant on each component of K-L (i.e. if U is a component of $|K|-|L|$, then $c(v_0) = c(v_1)$ for every vertex of U; for let $\langle v_0, v_1\rangle$ be a polygonal arc (1-chain) from v_0 to v_1 in U, since $\delta c(\langle v_0, v_1\rangle) = 0$ we have $0 = \delta c(\langle v_0, v_1\rangle) = c(\partial\langle v_0, v_1\rangle) = c(v_1) - c(v_0))$. Suppose we have 0-cochains $c_1, \ldots, c_n$ representing linearly independent elements of H_e^0. Then there is a finite complex L such that each component of $|K|-|L|$ is infinite and each c_1 is a constant on each component, i.e., there exists L_i such that $\delta c_i = 0$ on K–L_i, then $L = \cup L_i$ enlarged as above. Since c_i's are linearly independent elements in $H_e^0(K)$ we have that $n \leqslant n(L) \leqslant e(K)$. Therefore $\dim H_e^0 \leqslant e(K)$.

Now suppose we remove a finite subcomplex L, leaving $n(L)$ infinite components. We construct $n(L)$ different 0-cochains, each being 1 on one component and 0 elsewhere. These cochains define linearly independent elements of H_e^0. Hence $n(L) \leqslant \dim H_e^0$ and so $e(K) \leqslant \dim H_e^0$ and the lemma is established.

Proof of Proposition I.7: The only facts that are needed about M are that $H_{n-1}(M) = 0$, M is connected, not compact and has no boundary. For then by the Poincaré Duality Theorem $H_f^1(M) \cong H_{n-1}(M) = 0$.

Hence we get the following short exact sequence:

$$0 \longrightarrow H_f^0(M) \longrightarrow \underset{\mathbf{Z}_2}{\overset{}{\underset{\wr\wr}{H^0(M)}}} \longrightarrow H_e^0(M) \longrightarrow 0 \; (= H_f^1(M))$$

Claim $H_f^0(M) = 0$. Then this implies $H_e^0(M) \approx \mathbf{Z}_2$ which by Lemma I.9 implies $\dim H_e^0 = e(M) = 1$. To show $H_f^0(M) = 0$, it will suffice to show that if c^0 is a 0-cocycle then $c^0 \equiv 0$. But if c^0 is a 0-cocycle then c^0 is constant on every vertex $v \in K$. But since $c^0 \in C_f^0(K)$, $c^0 =$ zero outside a finite number of vertices of K and since it is a constant, $c^0 \equiv 0$.

PROPOSITION I.10: *Let M^n be a contractible manifold which is 1-connected at ∞. Then for every compact $C \subset M$ there exists a compact D such that $C \subset D \subset M$ and $(M, M\text{-}D)$ is a 2-connected.*

Proof. Since M^n is 1-connected at ∞ there exists a compact D such that $M^n\text{-}D$ is 1-connected. Hence consider the following homotopy sequence:

$$\longrightarrow \pi_2(M^n, x_0) \xrightarrow{j*} \pi_2(M^n, M^n\text{-}D, x_0)$$
$$\xrightarrow{\partial *} \pi_1(M^n\text{-}D, x_0) \xrightarrow{i*} \pi_1(M^n, x_0).$$

Since $\pi_2(M^n, x_0) = 0$ and $\pi_1(M^n\text{-}D, x_0) = 0$, then it follows that $\pi_2(M^n, M^n\text{-}D, x_0) = 0$. Also since $M^n\text{-}D$ is 1-connected and $\pi_1(M^n) = 0$, it follows that $\pi_1(M^n, M^n\text{-}D, x_0) = 0$ also.

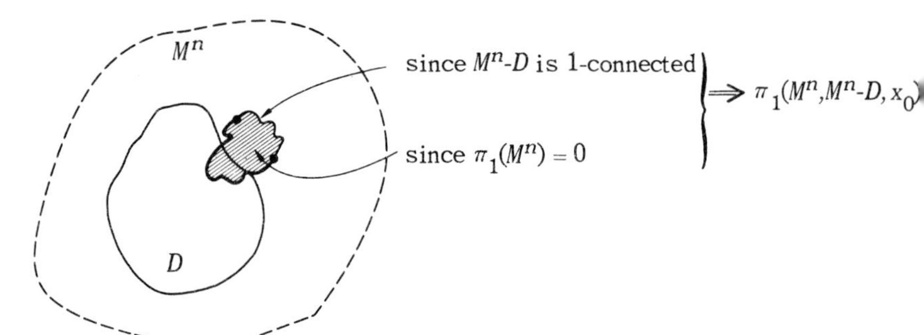

We now return to obtaining our main results and doing what this chapter promised, namely applying the engulfing lemma.

PROPOSITION I.11: *Suppose M^n is a contractible pwl n-manifold without boundary ($n \geqslant 5$) that is 1-connected at ∞, C is a compact subset of M^n and T^2 = 2-skeleton of the pwl triangulation of M^n. Then there exists a compact set $E_1 \subset M^n$ and a pwl homeomorphism h_1 of M^n onto itself such that $C \subset E_1 \subset M$, $T^2 \subset h_1(M\text{-}C)$ and* $h_1 \mid M\text{-}E_1$ = identity.
(i.e. We can move any compact set off the 2-skeleton of M^n).

Proof. Since M^n is 1-connected at ∞ there exists a compact D such that $C \subset D \subset M$ and $(M, M\text{-}D)$ is a 2-connected (by Proposition I.10). Let $U = M\text{-}D$ and $P = T^2$.

Now $P \cap (M-U) = P \cap D$ is compact, hence by the engulfing lemma there exists a compact set E and a pwl homeomorphism

$h_1\colon M \to M$ such that $h_1 \mid M-E$ = identity, $T^2 \subset h_1(M\text{-}D) \subset h_1(M\text{-}C)$. The E_1 above is $E \cup D$.

THEOREM I.1: *Let M^n be a contractible pwl manifold without boundary which is 1-connected at ∞. If $n \geqslant 5$, then M is pwl homeomorphic to E^n.*

Proof. First we will show (A): If C is compact $\subset M$, then there exists a pwl ball F such that $C \subset \text{int } F \subset M$. Now given C compact, applying Proposition I.11, there exists $h_1\colon M \to M$ and E_1 compact such that $T^2 \subset h_1(M\text{-}C)$ and $h_1 \mid M\text{-}E_1$ = identity. Let $K = \mathrm{T}^2 +$ {all closed simplexes of T (the triangulation of M) which miss $h_1(C)$}. Let L = {subcomplex of T' maximal with respect to missing K}. L is compact, misses T^2, dim $L \leqslant n\text{-}3$ there exists a pwl n-ball $A \subset M$ such that $L \subset$ int A (by Corollary IV.18 of Volume I). Now we make use of the fact that every n-simplex of M is the join of a simplex of K' and a simplex of L. That is we have $|L| \subset$ int A and $|K'| \subset h_1(M\text{-}C)$ and applying Lemma IV.12 of Volume I, there exists $h_2\colon M \to M$ such that $h_2(\text{int } A) \cup h_1(M\text{-}C) = M$. That is, $h_1(C) \subset h_2(\text{int } A)$. Then the desired n-ball F is $h_1^{-1}\, h_2(A)$.

Using (A) we can show (B): M is the union of a sequence $\{F_i\}$ of pwl n-balls such that $F_i \subset \text{int } F_{i+1}$. That is let $\{\Delta_i\}$ be the n-simplexes of M^n giving the locally finite triangulation of M^n. By (A) there exists F_1 a pwl ball such that $\Delta_1 \subset$ int F_1. Considering C of (A) now as $F_1 \cup \Delta_2$ there exists F_2 a pwl ball such that $F_1 \cup \Delta_2 \subset$ int F_2. In general taking C of (A) as $\mathrm{F}_{i-1} \cup \Delta_i$ there exists F_i such that $F_{i-1} \cup \Delta_i \subset$ int F_i. Since $\cup \Delta_i = M^n$, we have $M^n = \cup F_i$ and each $F_i \subset$ int F_{i+1}.

With these remarks now complete we finish the proof. Now certainly $E^n = \cup G_i$, where the G_i are pwl n-balls such that $G_i \subset \text{int } G_{i+1}$ for all i. Suppose inductively, we have a pwl homeomorphism $h_i\colon F_i \to G_i$ such that $h_i \mid F_j$ takes F_j pwl onto G_j and $h_i \mid F_j = h_j \mid F_j$, $j \leqslant i$. Now there exists a pwl homeomorphism f taking $F_{i+1} \to G_{i+1}$ such that $h_i f^{-1} \mid f(F_i)$ is orientation preserving (may have to reflect if necessary). Each of $f(F_i)$ and $h_i(F_i) = G_i$ are pwl balls of G_{i+1} and $h_i f^{-1}$ is a + PLO taking $f(F_i)$ onto G_i. By the result of section D of Chapter IV of Volume I there exists g a PLO taking $G_{i+1} \to G_{i+1}$ such that $g \mid f(F_i) = h_i f^{-1}$. Hence let h_{i+1} be the extension of h_i given by $h_{i+1} = g \circ f$. Then $h_{i+1} \mid F_i = gf \mid F_i = g \mid f(F_i) = h_i f^{-1} \mid f(F_i) = h_i(F_i)$. Thus it follows that M^n is combinatorially equivalent to E^n.

COROLLARY I.2. *E^n has a unique pwl structure for $n \geqslant 5$.*

Proof: If M is any combinatorial n-manifold which is topologically homeomorphic to E^n, then it is contractible and 1-connected at ∞. Therefore by Theorem I.1, M is pwl homeomorphic to E^n. (In fact, shortly we will see that for any vertex $v \in E^n$ there exists a pwl homeomorphism $h\colon \overset{\circ}{\mathrm{st}}(v, E^n) \xrightarrow{\text{onto}} E^n$ which approximates a radial expansion).

COROLLARY I.3. *If X and Y are contractible pwl manifolds without boundary such that $\dim X \geqslant 1$, $\dim Y \geqslant 1$ and $\dim X + \dim Y = n \geqslant 5$, then $X \times Y$ is homeomorphic to E^n.*

Proof: By Proposition I.8, $X \times Y$ is 1-connected at ∞ and by Theorem I.1, $X \times Y$ is $\approx E^n$.

COROLLARY I.4. *If M^n is a homotopy pwl n-sphere, $n \geqslant 5$ then $M^n - \{p\}$ is pwl homeomorphic to E^n and hence M^n is topologically equivalent to S^n.*

Proof: Let σ be an n-simplex of M^n and $p \in \operatorname{int} \sigma$. $M^n - \{p\}$ is contractible and, clearly, 1-connected at ∞ (for just take a small open n-cell about p in the simplex σ missing any given compact set). Finally, we want to make $M^n - \{p\}$ combinatorial – so triangulate 'toward' p (refer to diagram below!). Hence by Theorem I.1, $M^n - \{p\} \approx E^n$ and M^n is topologically equivalent to S^n.

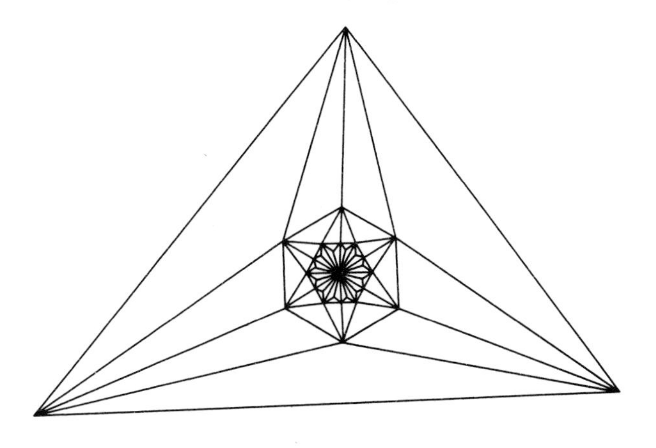

§B. Approximating Stable Homeomorphisms by Piecewise Linear Ones

The ideas of this section are based on the paper by E.H. Connell, 'Approximating stable homeomorphisms by piecewise linear ones', *Annals of Math.*, 78 (1963), 326–338. The discussion here will include some results assumed in that paper and a modification of some of the proofs given here. First we give some definitions.

A homeomorphism (not necessarily pwl) h of E^n or S^n onto itself is stable if there exist homeomorphisms $h_1, h_2, \ldots, h_m$ and non-void open sets $U_1, U_2, \ldots, U_m$ such that $h = h_m \circ h_{m-1} \circ \ldots \circ h_1$ and $h_i \mid U_i$ = identity for $n = 1, 2, \ldots, m$. It is known that all orientation preserving homeomorphisms of E^n or S^n are stable provided $n = 1, 2$ or 3. For $n \geqslant 4$, this is an unsolved problem and the conjecture that all orientation-preserving homeomorphisms of E^n or S^n are stable is equivalent to the annulus conjecture. We recall that in section D of Chapter IV of Volume I we showed that all pwl orientation-preserving homeomorphisms of S^n onto itself are isotopic to the identity. It also follows easily from the techniques developed there that any pwl orientation-preserving homeomorphism of E^n onto E^n is stable.

Let $H(S^n)$ (or $H(E^n)$) denote the group of all homeomorphisms of S^n (or E^n) onto itself where multiplication is composition. Let $SH(S^n)$ (or $SH(E^n)$) denote the subgroup of stable homeomorphisms. $SH(S^n)$ is a normal subgroup of $H(S^n)$ and in fact we will see shortly that $SH(S^n) = \{$intersection of all normal (nontrivial) subgroups of $H(S^n)\}$ and is simple.

Exercise I.1. Any stable homeomorphism is isotopic to the identity by an isotopy h_t such that h_t is stable for every t, $0 \leqslant t \leqslant 1$.

The main results are as follows:

THEOREM I.12: *Let T be an arbitrary pwl triangulation on S^n $(n \geqslant 7)$ and let $h: S^n \to S^n$ be a stable homeomorphism. If $\epsilon > 0$, then there exists a pwl homeomorphism (relative to T) $f: S^n \to S^n$ such that $|h(x) - f(x)| < \epsilon$ for all $x \in S^n$.*

THEOREM I.13. *Let T be an arbitrary pwl triangulation on E^n $(n \geqslant 7)$. If $h: E^n \to E^n$ is a stable homeomorphism and $\epsilon(x): E^n \to (0, \infty)$ is a continuous function, then there exists a pwl homeomorphism (relative to T) $f: E^n \to E^n$ such that $|f(x) - h(x)| < \epsilon(x)$ for $x \in E^n$.*

Remark: The lemmas necessary to obtain the result for $n \geqslant 7$ have been extended by R.H. Bing so as to give the above theorems for $n \geqslant 5$. This involves the same type of trick used in getting the engulfing lemma for $p = n-3$ plus additional delicate epsilontics and careful constructions.

LEMMA I.14. *Let $h \in H(S^n)$ such that $h \neq$ identity and $f \in SH(S^n)$ such that $f =$ identity on some $V^{\text{open}} \subset S^n$, then $f =$ product of 4 conjugates of h^{-1} and h by homeomorphisms in $SH(S^n)$ (in fact a total of 20 homeomorphisms). In particular if f is any arbitrary homeomorphism in $SH(S^n)$ (i.e. $f = f_k \circ f_{k-1} \circ \cdots \circ f_1$, where for each i there exists $V_i^{\text{open}} \subset S^n$ such that $f_i|_{V_i} =$ identity), then f is the product of $4k$ conjugates of h^{-1} and h by homeomorphisms in $SH(S^n)$.*

Remark: It follows from this that $SH(S^n) = \cap N$ of all nontrivial normal subgroups of $H(S^n)$ and furthermore, $SH(S^n)$ is simple. That is, since $SH(S^n)$ is normal, $\cap N \subset SH(S^n)$. Let $h \in N$ such that $h \neq$ identity, then given any $f \in SH(S^n)$, $f =$ product of $4k$ conjugates of h^{-1} and h by homeomorphisms in $SH(S^n)$. That is, $f_k = (g_{k_1}^{-1} h^{-1} g_{k_1})(g_{k_2}^{-1} h g_{k_2})(g_{k_3}^{-1} h^{-1} g_{k_3})(g_{k_4}^{-1} h g_{k_4})$ where $g_{k_i} \in SH(S^n)$ $i = 1, 2, 3, 4$. Therefore $f_k \in N$ and hence $f = f_k \circ f_{k-1} \circ \ldots \circ f_1 \in N$. Thus $SH(S^n) \subset$ each N. Similarly, suppose $A \neq \{\text{identity}\}$ is a normal subgroup of $SH(S^n)$. Let $h \in A$ such that $h \neq$ identity and $f \in SH(S^n)$, then as above $f \in A$. Therefore $SH(S^n) \subset A$ and hence $SH(S^n)$ is simple.

Proof: Let $h \in H(S^n)$ such that $h \neq$ identity. Suppose $f \in SH(S^n)$ such that $f \neq$ identity on V. Since $h \neq$ identity there exists open set $U \subset S^n$ such that $U \cap h^{-1}(U) = \emptyset$. Let N be a n-cell in U and $N_0, N_1, N_2, \ldots$ a sequence of disjoint n-cells in int N converging to a point of int N. Let r' be a homeomorphism of S^n onto itself such that $r' =$ identity on $S^n\text{-}N$ and $r'(N_i) = N_{i+1}$, $i = 0, 1, 2, \ldots$.

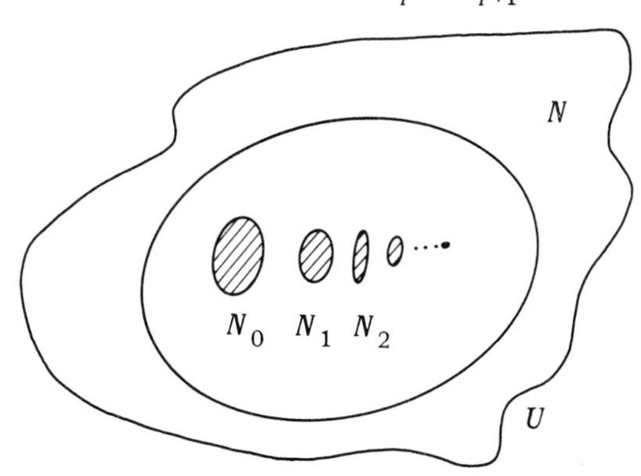

Define $r = r'^{-1}$. Hence $r(N_{i+1}) = N_i$, $r(N_1) = N_0$, $r^2(N_2) = N_0, \ldots,$ $r^i(N_i) = N_0$. Suppose $g \in SH(S^n)$ such that $g =$ identity outside of N_0, then $r^{-i} g r^i =$ identity outside of N_i. Define φ taking S^n onto itself by $\phi|_{N_i} = r^{-i} g r^i$, ϕ = identity on $S^n - \cup_i N_i$ and $\phi|_{N_0} = g|_{N_0}$. Note $\phi|_{N_i} : N_i \to N_i$. We now claim that $(r^{-1}\phi^{-1}h^{-1}\phi r)(r^{-1}hr)h^{-1}(\phi^{-1}h\phi) = g$. Let us denote this by $(*)$. First the left side $= (r^{-1}\phi^{-1}h^{-1}\phi h)\, r(h^{-1}\phi^{-1}h)\,\phi$. Also $r(h^{-1}\phi^{-1}h) = (h^{-1}\phi^{-1}h)r$. This follows since ϕ = identity outside N, and hence $h^{-1}\phi^{-1}h$ = identity outside $h^{-1}(N)$. Since r = identity outside N, the fact that $N \cap h^{-1}(N) = \emptyset$ gives the above. Thus $(r^{-1}\phi^{-1}h^{-1}\phi h)(h^{-1}\phi^{-1}h)\, r\phi = r^{-1}\phi^{-1}r\phi = g$. For ϕ = identity on $S^n - \cup N_i$ and hence $r^{-1}\phi^{-1}r\phi$ = identity on $S^n - \cup N_i$, $r^{-1}\phi^{-1}r\phi|_{N_0} = g|_{N_0}$ and $r^{-1}\phi^{-1}r\phi|_{N_i}$ = identity. Therefore $r^{-1}\phi^{-1}r\phi = g$.

Now let $t \in SH(S^n)$ such that $t(S^n - V) \subset N_0$. Then tft^{-1} = identity on $t(V)$ and this implies tft^{-1} = identity on $S^n - N_0$. We let $tft^{-1} = g$ as used above. Now by $(*)$, $tft^{-1} = (r^{-1}\phi^{-1}h^{-1}\phi r)(r^{-1}hr)h^{-1}(\phi^{-1}h\phi)$ and hence $f = (t^{-1}r^{-1}\phi^{-1}h^{-1}\phi rt)(t^{-1}r^{-1}hrt)(t^{-1}ht)(t^{-1}\phi^{-1}h\phi t)$ and the lemma follows.

Notation: Let $\mathcal{O}^n$ denote the open unit ball in E^n. For $t \in (0, \infty)$ let $t\mathcal{O}^n$ denote open ball in E^n of radius t centred at the origin. If $\{0\}$ is the origin of E^n and $x \neq \{0\} \neq y$, then $\theta\{x, y\}$ will represent the angle in radians between the two line intervals, one joining $\{0\}$ to x and the other joining $\{0\}$ to y.

LEMMA I.15. *Suppose $E^n (n \geqslant 4)$ has an arbitrary pwl structure T, K is a finite subcomplex of T, $\dim K \leqslant n-4$, a, b and ϵ are numbers with $0 < a < b$ and $\epsilon > 0$, and $K \subset b\mathcal{O} = b\mathcal{O}^n$. Then there exists a pwl homeomorphism (relative to T) carrying E^n onto E^n such that $h|_{(a-\epsilon)\mathcal{O}}$ = identity, $h|_{E^n - b\mathcal{O}}$ = identity, $h(a\mathcal{O}) \subset K$ and $\theta\{h(x), x\} < \epsilon$ for $x \in E^n$.*

Proof: Let n be fixed, $n \geqslant 4$. The proof will be by induction on $\dim K$. The lemma is immediate for $\dim K = 0$. Assume the lemma is true for dim $K \leqslant m$, $m \leqslant n$-5. Now let K, T, a, b, and ϵ be given, dim $K = m+1$, $0<a<b$, $\epsilon>0$ and let $\epsilon<a$. Since we will be leaving points fixed in $(a-\epsilon)\mathcal{O}$, we can assume without loss of generality that the origin $0 \notin K$. Let T_1 be a subdivision of T such that each closed simplex $\sigma \in T_1$ which intersects $(b+\epsilon)\mathcal{O}$ has diameter $<\epsilon/5$ and if σ intersects E^n -$(a-\epsilon)\mathcal{O}$ and $x, y \in \sigma$, then $\theta\{x,y\} < \epsilon/5$. Let K_1 be K under the new triangulation T_1.

Let us consider the complex $K_1 \times [0,1]$ and identify $K_1 \times \{0\}$ with K_1. Define f_1: $K_1 \times [0,1] \rightarrow b\mathcal{O} \subset E^n$ as follows: $f_1(k,0) = k$, $f_1(k,1) = \{(a-\epsilon/2)/\,||k||\,\}k$ and $f_1(k,t) = tf_1(k,1) + (1-t)f_1(k,0)$. Let $f: K_1 \times [0,1] \rightarrow E^n$ be a pwl general position approximation to f_1 such that $f(k,0) = f_1(k,0) = k$ for $k \in K_1$. Let $K_2 \times [0,1]$ and T_2 be subdivisions of $K_1 \times [0,1]$ and T_1 respectively so that f is simplicial. Let the approximation be so close that f has the following properties:

(1) If $\sigma \in K_1 \subset T_1$ and $(x,t) \in \sigma \times [0,1] \subset K_1 \times [0,1]$ and $(y,s) \in \sigma \times [0,1]$ then $\theta\{f(x,t), f(y,s)\} < \epsilon/5$.

(2) $f(K_1 \times [0,1]) \subset b\mathcal{O}$ and $f(K_1 \times \{1\}) \subset (a-2\epsilon/5)\mathcal{O} \cap [E^n\text{-}\overline{(a-3\epsilon/5)\mathcal{O}}]$.

(3) If $\sigma \in K_1 \subset T_1$ and $\sigma \subset (a-\epsilon/5)\mathcal{O}$, then $f(\sigma \times [0,1]) \subset (a-\epsilon/5)\mathcal{O}$.

(4) If $\sigma \in K_1 \subset T_1$, $\sigma \subset E^n\text{-}\overline{(a-3\epsilon/5)}\mathcal{O}$, then $f(\sigma \times [0,1]) \subset E^n\text{-}\overline{(a-3\epsilon/5)}\mathcal{O}$.

(5) If $\sigma \in K_1 \subset T_1$, $\sigma \subset E^n\text{-}\overline{(a-\epsilon)\mathcal{O}}$, then $f(\sigma \times [0,1]) \subset E^n\text{-}\overline{(a-\epsilon)\mathcal{O}}$.

Let S denote Sing f and $K_3 \times [0,1]$ and T_3 be subdivisions of $K_2 \times [0,1]$ and T_2 respectively such that S is a subcomplex of $K_3 \times [0,1]$ and $f(S)$ is a subcomplex of T_3. Now dim $S =$ dim $f(S) \leqslant 2(m+2) - n \leqslant 2(m+2) - (m+5) = m-1$. Let L_1 be the $(m-1)$-skeleton of $K_3 \times \{0\}$ and $L = (L_1 \times [0,1]) \cup (S \times [0,1])$. Let $K_4 \times [0,1]$ and T_4 be subdivisions so that L is a subcomplex of $K_4 \times [0,1]$ and $f(L)$ is a subcomplex of T_4. Since dim $L =$ dim $f(L) \leqslant m$, by induction there exists a pwl homeomorphism

$g_1 \colon E^n \to E^n$ such that:

(i) $g_1 | (a-\epsilon/5)\mathcal{O} =$ identity and $g_1 |_{E^n - b\mathcal{O}} =$ identity,

(ii) $g_1(a\mathcal{O}) \subset f(L)$,

and (iii) $\theta\{g_1(x), x\} < \epsilon/5$ for $x \in E^n$.

Let $s_1, s_2, \ldots, s_p$ be the open m simplexes of $K_3 \times \{0\}$ such that $f(s_i) \not\subset (a-2\epsilon/5)\mathcal{O}$. Let $u_1, u_2, \ldots u_p$ be polyhedra in $s_1, s_2, \ldots, s_p$ respectively so that $(S \times [0,1]) \cap (s_i \times [0,1]) \subset (s_i - u_i) \times [0,1]$ and $f((s_i - \text{int } u_i) \times [0,1]) \subset g_1(a\mathcal{O})$, $i = 1, 2, \ldots, p$. Then f is a pwl embedding on $\cup_{i=1}^{p} u_i \times [0,1]$. Let $U_1, U_2, \ldots, U_p$ be a disjoint open sets in $b\mathcal{O} - \overline{(a-\epsilon)\mathcal{O}}$ such that $f(u_i \times [0,1]) \subset U_i$, $U_i \subset \epsilon/5$-neighborhood of $f(u_i \times [0,1])$ and if $x_i, y_i \in U_i$, then $\theta\{x_i, y_i\} < \epsilon/5$. Let $K_4 \times [0,1]$ and T_5 be further subdivisions so that $u_i \times [0,1]$ is a subcomplex of $K_5 \times [0,1]$ for each i, so that $u_i \times [0,1] \searrow u_i \times \{1\} \cup (u_i - \text{int } u_i) \times [0,1]$, so that $f(u_i \times [0,1])$ is a subcomplex of T_5, and so that for every $\sigma \in f(u_i \times [0,1])$ we have $st(\sigma, T_5) \subset U_i$. Since f is an embedding on each $u_i \times [0,1]$, we have induced collapse $f(u_i \times [0,1]) \searrow f(u_i \times \{1\} \cup (u_i - \text{int } u_i) \times [0,1]) \subset g_1(a\mathcal{O})$. Hence there exists pwl homeomorphism $g_2 \colon E^n \to E^n$ so that $g_2(g_1(a\mathcal{O})) \supset f(L \cup K_3^{(m)} \times [0,1])$, where $K_3^{(m)}$ is the m-skeleton of K_3. Also $g_2 =$ identity outside $\cup_{i=1}^{p} U_i$ and hence $\theta\{g_2(x), x\} < \epsilon/5$.

Similarly if $t_1, t_2, \ldots, t_g$ are the open $(m+1)$-simplexes of $K_3 \times \{0\}$ we can get polyhedra as above in each of these, so that f on the polyhedra $\times [0,1]$ is an embedding. Also as above, we obtain a collection of disjoint open subsets of $b\mathcal{O}$ having properties similar to the above. Using the same techniques there exists pwl homeomorphism $g_3 \colon E^n \to E^n$ so that $g_3(g_2(g_1(a\mathcal{O}))) \supset f(L \cup K_3 \times [0,1])$, $g_3 =$ identity outside the collection of disjoint open sets and $\theta\{g_3(x), x\} < \epsilon/5$.

Since each of f, g_1, g_2, g_3 had an 'angular approximation' $< \epsilon/5$, clearly $h = g_3 \circ g_2 \circ g_1$ is the desired pwl homeomorphism. That is, $h|_{(a-\epsilon)\mathcal{O}} =$ identity, $h|_{E^n - b\mathcal{O}} =$ idendity, $h(a\mathcal{O}) \supset K$ and $\theta\{h(x), x\} < 4\epsilon/5 < \epsilon$ for $x \in E^n$.

LEMMA I.16. *Suppose E^n $(n \geqslant 7)$ has an arbitrary pwl structure T, and a, b and ϵ are numbers with $0 < a < b$ and $\epsilon > 0$. Then there exists a homeomorphism h: $E^n \to E^n$ such that h is pwl relative to T, $h|_{(a-\epsilon)\mathcal{O}} =$ identity, $h|_{E^n (b+\epsilon)\mathcal{O}} =$ identity, $h(a\mathcal{O}) \supset b\mathcal{O}$ and $\theta\{h(x), x\} < \epsilon$ for $x \in E^n$.*

Proof: Let us subdivide T sufficently fine - say to $\hat{T}$, such that if $\sigma \cap [\overline{(b+\epsilon/2)\mathcal{O}} - (a-\epsilon/2)\mathcal{O}] \neq \emptyset$, then $\sigma \cap (E^n - (b+\epsilon)\mathcal{O}) = \emptyset$ and $\sigma \cap (a-\epsilon)\mathcal{O} = \emptyset$. Let $K = \cup\sigma$ plus their faces such that $\sigma \in \hat{T}$ and $\sigma \cap [\overline{(b+\epsilon/2)\mathcal{O}} - (a-\epsilon/2)\mathcal{O}] \neq \emptyset$. Also let us suppose that $N(K, \hat{T}) \subset (b+\epsilon)\mathcal{O} - \overline{(a-\epsilon)\mathcal{O}}$ and each $\sigma \in N(K, \hat{T})$ has an angular diameter $< \epsilon/3$. Let K^3 be the 3-skeleton of K and $\hat{K} = K^3 \cup \overline{(\hat{T} \cap a\mathcal{O}) - K}$. Let $\hat{L}$ be the complementary complex of $\hat{K}$ in $\hat{T}$ (i.e., $\hat{L} =$ subcomplex of $\hat{T}'$ maximal with respect to missing $\hat{K}'$). Then dim $(\hat{L} \cap (b+\epsilon/2)\mathcal{O}) = n-4$ and each n-simplex of $\hat{T}'$ is either in $\hat{K}'$ or $\hat{L}$ or is the join of a simplex in $\hat{K}'$ and a simplex in $\hat{L}$. Let $L = (n-4)$-skeleton of the subcomplex of $\hat{L}$ made up of those simplexes of $\hat{L}$ intersecting $\overline{(b+\epsilon/2)}\mathcal{O}$ plus their faces. Now $|\hat{L} \cap (b+\epsilon/2)\mathcal{O}| \subset |L|$ and $E^n - (b+\epsilon)\mathcal{O} \subset |\hat{L}|$.

By Lemma I.15, there exists h_1: $E^n \to E^n$ such that $h_1 =$ identity on $(a-\epsilon/2)\mathcal{O}$ and on $E^n - (b+\epsilon)\mathcal{O}$, $h_1(a\mathcal{O}) \supset K^3$ (here we use the fact that $n \geqslant 7$) and $\theta\{h_1(x), x\} < \epsilon/3$. Hence h_1 expands $a\mathcal{O}$ outward pwl approximating a radial expansion. Now we want a pwl contraction approximating a radial contraction engulfing L and leaving points fixed on $(a-\epsilon)\mathcal{O}$ and on $E^n - (b+\epsilon/2)\mathcal{O}$. Using the same techniques as in Lemma I.15, we can obtain a h_2: $E^n \to E^n$

such that h_2 = identity on $(a-\epsilon)\mathcal{O}$ and on E^n-$(b+\epsilon/2)\mathcal{O}$ such that $h_2(E^n\text{-}\overline{(b+\epsilon/4)\mathcal{O}}) \supset L$ and $\theta\{h_2^{-1}(x), x\} < \epsilon/3$. Now we have $\hat{K}' \subset h_1(a\mathcal{O})$ and $\hat{L} \subset h_2(E^n\text{-}\overline{(b+\epsilon/4)}\mathcal{O})$ and by the definitions of $\hat{L}$ and $\hat{K}'$ we can obtain a pwl homeomorphism h_3: $E^n \to E^n$ such that h_3 = identity on $\hat{K}'$ and $\hat{L}$ and hence $(a-\epsilon)\mathcal{O}$ and E^n-$(b+\epsilon)\mathcal{O}$ and such that $h_3 h_1(a\mathcal{O}) \cup h_2(E^n\text{-}\overline{(b+\epsilon/4)}\mathcal{O}) = E^n$. h_3 only moves points in $(b+\epsilon)\mathcal{O}$-$\overline{(a-\epsilon)}\mathcal{O}$ and since $h_3(\sigma) = \sigma$ for any simplex of $\hat{T}'$ and each simplex of $\hat{T} \cap N(K, \hat{T})$ has angular diameter $<\epsilon/3$, it follows that $\theta\{h_3(x), x\} < \epsilon/3$.

Let $h = h_2^{-1} \circ h_3 \circ h_1$. Then h = identity on $(a-\epsilon)\mathcal{O}$ and on E^n-$(b+\epsilon)\mathcal{O}$ and $\theta\{h(x), x\} < \epsilon$. Since $h_3 h_1(a\mathcal{O}) \cup h_2(E^n\text{-}\overline{(b+\epsilon/4)\mathcal{O}}) = E^n$, $h_2^{-1} \circ h_3 \circ h_1(a\mathcal{O}) \cup (E^n\text{-}\overline{(b+\epsilon/4)\mathcal{O}}) = E^n$ and therefore $h(a\mathcal{O}) \supset b\mathcal{O}$.

Let G_n be the subgroup under composition of all homeomorphisms $h \in H(S^n)$ such that, if T is any pwl structure on S^n and given any $\epsilon > 0$ there exists a homeomorphism $f \in H(S^n)$ such that f is pwl relative to T and $|h(x) - f(x)| < \epsilon$ for all $x \in S^n$.

LEMMA I.17. *G_n is a normal subgroup of $H(S^n)$.*

Proof: Suppose $h \in G_n$ and $g \in H(S^n)$, we want to show $g^{-1}hg \in G_n$. Let T and $\epsilon > 0$ be given. There exists a $\delta > 0$ such that if $|x-y| < \delta$ then $|g^{-1}(x) - g^{-1}(y)| < \epsilon$. Let T_1 be the pwl structure on S^n which is the image of T under g, i.e. $T_1 = g(T)$. Since $h \in G_n$, there exists $f \in H(S^n)$ such that f is pwl relative to T_1 and so that $|h(x) - f(x)| < \delta$ for $x \in S^n$. Thus $|g^{-1}hg(x) - g^{-1}fg(x)| < \epsilon$ for $x \in S^n$ and $g^{-1}fg$ is pwl relative to T, since g is pwl from T to T_1, f is pwl from T_1 to T_1 and g^{-1} is pwl from T_1 to T.

THEOREM I.12. *Let T be an arbitrary pwl structure on S^n $(n \geqslant 7)$ and h: $S^n \to S^n$ be a stable homeomorphism. If $\epsilon > 0$, then there exists a homeomorphism f: $S^n \to S^n$ such that f is pwl relative to T*

and $|h(x)-f(x)|<\epsilon$ *for* $x \in S^n$.

Proof: By Lemma I.14, $SH(S^n)$ is a simple normal subgroup of $H(S^n)$, hence if we show that $SH(S^n) \cap G_n$ contains more than the identity, it will follow that $SH(S^n) \subset G_n$. (Also by Lemma I.14 we know $SH(S^n) = \cap N$, N normal nontrivial subgroup of $H(S^n)$ so it actually suffices to show that there exists an $h \neq$ identity $\in G_n$, but the one we will construct belongs to $SH(S^n)$ anyway). Let u be a homeomorphism taking $[0,1] \to [0,1]$ such that $u(0) = 0$, $u(1) = 1$ and if $0<r<1$ then $r<u(r)<1$. Let h be the homeomorphism of E^n onto E^n such that $h(x) = x$ for $||x|| \geqslant 1$, $h(0) = 0$, $\theta\{h(x), x\} = 0$ for every x, and $h(x) = u(||x||)x/||x||$ for $0<||x|| \leqslant 1$. Let T be any pwl structure on E^n and $\epsilon>0$. It will be shown that there exists f: $E^n \to E^n$ which is a pwl homeomorphism relative to T so that $f(x) = x$ for $||x|| \geqslant 1$ and $|h(x)-f(x)|<\epsilon$ for $x \in E^n$. Since that of h and f then extend to a homeomorphism from S^n to S^n by defining $h(\infty) = f(\infty) = \infty$, this will complete the proof of Theorem V.12.

Let $0 = r_0 < r_1 < r_2 < \ldots < r_{m+1} = 1$ be numbers such that $(u(r_{i+2})-u(r_i)) < \epsilon/2$ for $i = 0, 1, 2, \ldots, m-1$. By Lemma I.16 there exists pwl homeomorphisms $f_1, f_2, \ldots, f_m$ such that $f_i|_{r_{i-1}\mathcal{O}} =$ identity, $f_i|_{E^n - u(r_{i+1})\mathcal{O}} =$ identity, $\theta\{f_i(x), x\} < \epsilon/4$ for $x \in E^n$, and $f_i(r_i\mathcal{O}) \supset u(r_i)\mathcal{O}$. Let $f = f_1 \circ f_2 \circ \ldots \circ f_m$. Now f is a homeomorphism of E^n onto E^n that is pwl relative to T and $f|_{E^n - \mathcal{O}} =$ identity. The claim is that $|f(x)-h(x)|<\epsilon$ for $x \in \mathcal{O}$. Let $x \in r_{i+1}\mathcal{O} \cap (E^n - r_i\mathcal{O}) = r_{i+1}\mathcal{O} - r_i\mathcal{O}$, $0 \leqslant i \leqslant m$; then $f(x) = f_i \circ f_{i+1}(x)$, because $f_t|_{r_{i+1}\mathcal{O}} =$ identity for $t > i+1$, $f_i(r_i\mathcal{O})$ contains $u(r_i)\mathcal{O}$ and $f_t|_{E^n - u(r_i)\mathcal{O}} =$ identity for $t<i$. Now $h(x) \in u(r_{i+1})\mathcal{O} - u(r_i)\mathcal{O}$ and $f(x) \in u(r_{i+2})\mathcal{O} - u(r_i)\mathcal{O}$, therefore $||h(x)||$ and $||f(x)||$ differ by $<\epsilon/2$. Since $\theta\{h(x), f(x)\} < \epsilon/2$ also, $|h(x)-f(x)|<\epsilon$.

LEMMA I.18. *Let r be a homeomorphism carrying* $[0,1)$ onto $[0,\infty)$ *so that* $r(0) = 0$ *and* $r(t) > t$ *for* $t \in (0,1)$. *Let g be the homeomorphism carrying* $\mathcal{O}^n$ *onto* E^n *defined as* $g(0) = 0$ *and for* $x \in \mathcal{O}$ *with* $\|x\| \neq 0$, $g(x) = r(\|x\|)\, x/\|x\|$. *Suppose* E^n $(n \geqslant 7)$ *has an arbitrary pwl structure T and* $\epsilon(x)$ *is a positive continuous function defined on* E^n, *then there exists* $F : \mathcal{O} \to E^n$ *such that F is pwl with respect to T and* $|Fg^{-1}(x) - x| < \epsilon(x)$.

Proof: Let $1 < r_1 < r_2 < \ldots$ be an increasing sequence of numbers tending to ∞ such that $r_i - r_{i-1} < \eta_i/4$, where $\eta_i = \min\{(1, \epsilon(x)) \mid x \in \overline{r_i \mathcal{O}}\}$. Let $s_i = r^{-1}(r_i)$. Also let $\delta_1, \delta_2, \ldots$ and $\gamma_1, \gamma_2, \ldots$ be a sequence of positive numbers so that for each i, $s_{i-1} < s_i - \delta_i < s_i$ and $r_i < r_i + \gamma_i < r_{i+1}$. Now for each i, let $f_i : E^n \to E^n$ be such that f_i is pwl with respect to T, $f_i =$ identity on $(s_i - \delta_i)\mathcal{O} \cup E^n - (r_i + \gamma_i)\mathcal{O}$, $f_i(s_i\mathcal{O}) \supset r_i\mathcal{O}$ and $\theta\{f_i(x), x\} < \eta_{i+1}/4\, r_{i+2}$.

Define $F : \mathcal{O} \to E^n$ by $F|_{s_i\mathcal{O}} = f_1 \circ f_2 \circ \ldots \circ f_i|_{s_i\mathcal{O}}$. Now if $x \in (s_i\mathcal{O} - s_{i-1}\mathcal{O})$, then $F(x) \in (r_{i+1} - r_{i-1})\mathcal{O}$. Hence $\big|\, \|F(x)\| - \|g(x)\| \,\big| < \eta_i/2$ and since $\theta\{F(x), x\} < \eta_i/2r_{i+1}$ and $\theta\{g(x), x\} = 0$, the angular distance between $g(x)$ and $F(x) < \eta_i/2$. Therefore $|g(x) - F(x)| < \eta_i < \epsilon(g(x))$.

THEOREM I.13. *Let T be an arbitrary pwl structure on* E^n $(n \geqslant 7)$. *If* $h : E^n \to E^n$ *is a stable homeomorphism and* $\epsilon(x) : E^n \to (0,\infty)$ *is a continuous function, then there exists a homeomorphism (pwl relative to T)* $f : E^n \to E^n$ *such that* $|f(x) - h(x)| < \epsilon(x)$ *for* $x \in E^n$.

Proof: Since h is stable, there exist homeomorphisms $h_1, \ldots, h_m$ and non-void open sets $U_1, \ldots, U_m$ such that $h_i|_{U_i} =$ identity. Hence if each h_i can be approximated by a pwl homeomorphism, then clearly h can also. Hence we may assume there is a non-void open set U such that $h|_U =$ identity. Also for convenience let us suppose that $\mathcal{O}^n \subset U$.

Let $h^{-1}(T) = T'$. Let $\eta(x)$ be a positive continuous function defined on E^n so that if $|y-x| < \eta(x)$ then $|h(y) - h(x)| < \epsilon(x)$. Let g be the function as defined in Lemma I.18, Then there exists a homeomorphism $F_1 : \mathcal{O}^n \to E^n$ that is pwl with respect to T and so that $|F_1(g^{-1}(x)) - x| < \eta(x)/2$. Similarly there exists a homeomorphism $F_2 : \mathcal{O}^n \to E^n$ that is pwl with respect to T' and so that $|F_2(g^{-1}(x)) - x| < \eta(x)/2$. Now $F_2 \circ F_1^{-1} : E^n \to E^n$ is a pwl homeomorphism from E^n under T to E^n under T' (since T $T | \mathcal{O}^n = T' | \mathcal{O}^n$) and $|F_2 \circ F_1^{-1}(x) - x| < \eta(x)$.

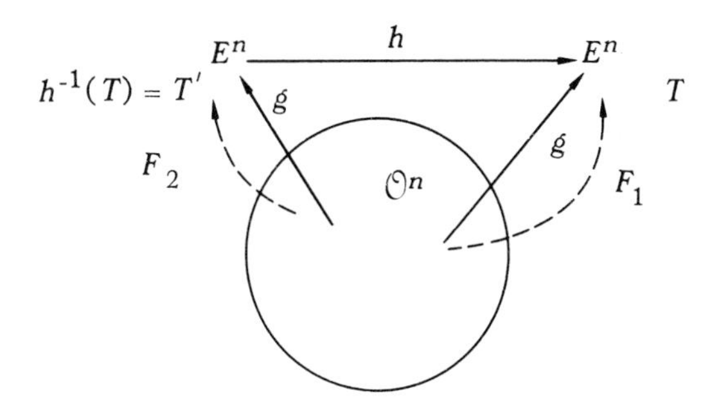

Let $f = h \circ F_2 \circ F_1^{-1} : E^n \to E^n$, then f is pwl relative to T and $|f(x) - h(x)| = |h \circ F_2 \circ F_1^{-1}(x) - h(x)| < \epsilon(x)$, since $|F_2 \circ F_1^{-1}(x) - x| < \eta(x)$.

THEOREM I.19. *Let T_1 and T_2 be two arbitrary pwl structures on E^n $(n \geqslant 7)$. If $g : E^n \to E^n$ is an orientation preserving homeomorphism pwl relative to T_1 and $\epsilon(x) : E^n \to (0, \infty)$ is continuous, then there exists a homeomorphism $f : E^n \to E^n$ such that f is pwl relative to T_2 and $|g(x) - f(x)| < \epsilon(x)$ for $x \in E^n$.*

Proof. This follows from Theorem V.13 after one observes that g is stable. To see that g is stable, let σ be an n-simplex of T_1 and let B be a pwl n-ball relative to T_1 containing $\sigma \cup g(\sigma)$ in its interior. Then by the results of Section D of Chapter IV of Volume I, there exists a pwl homeomorphism $\hat{h}$ (pwl relative to T_1) carrying B onto B such that $\hat{h}|_{\dot{B}}$ = identity and $\hat{h}|_{g(\sigma)} = g^{-1}$. Extend $\hat{h}$ to h taking $E^n \to E^n$ by defining h = identity on $E^n - B$. Then $g = h^{-1}(hg)$ and h^{-1} = identity on $E^n - B$ and $hg|_\sigma$ = identity and hence g is stable.

Exercise I.2.

Bounded homeomorphism on E^n $(n \geqslant 1)$ are stable. (i.e., if $f\colon E^n \to E^n$ is a homeomorphism and there exists an $M > 0$ such that $|f(x) - x| < M$ for $x \in E^n$, then f is stable.

Exercise I.3.

Suppose $g, h \in H(E^n)$ $(n \geqslant 1)$ and there exists $M > 0$ such that $|g(x) - h(x)| < M$ for all $x \in E^n$, then if $g \in SH(E^n)$ then $h \in SH(E^n)$.

§ C. *h*-Cobordism and the Hauptvermutung

A cobordism $(W;\, M, M')$ is a compact n-manifold W such that BdW is a disjoint union of M and M'. We will call $(W; M, M')$ a PL cobordism if W is also a combinatorial n-manifold. A PL cobordism $(W; M, M')$ is called a PL h-cobordism if $\pi_i\,(W, M) = 0 = \pi_i\,(W, M')$ for all i (equivalently, if both M and M' are deformation retracts of W). The Hauptvermutung is the conjecture that if two polyhedra are homeomorphic, then they are piecewise linearly homeomorphic and therefore have isomorphic simplicial subdivisions.

The main results of this section are (i) if $(W; M, M')$ is a PL h-cobordism, where dim $W = n \geqslant 5$, then $W - M' \approx M \times [0,1)$, (ii) for $n \geqslant 6$, there exist complexes K and L such that K is homeomorphic to L, but $K \not\approx L$, and (iii) there exists a sequence $\{K_i\}_{i=0}^{\infty}$ of PL $(n$-$2)$-spheres in S^n $(n \geqslant 6)$ such that the pairs $\{(S^n, K_i)\}_{i=0}^{\infty}$ are all topologically equivalent, but no two are PL homeomorphic.

Although our main concern in this course has been and is with results and techniques in combinatorial topology that are quite geometric in nature, a very important and powerful tool in attacking problems in combinatorial topology is a somewhat complicated algebraic structure, known as Whitehead torsion, that can be associated with a CW complex pair (K,L), where $|L|$ is a deformation retract of $|K|$. Because of its importance we will discuss it briefly here. We will not attempt to develop completely the algebraic machinery necessary to make the concept rigorous, for in order to obtain additional geometrical results, it will suffice here just to outline or state many of the results necessary to apply it in the geometrical setting. For further details one can refer to Milnor's lecture notes on Whitehead Torsion, Princeton, 1964 or to Milnor's article 'Whitehead Torsion', *Bull. Amer. Math. Soc.* 72(1966), 358–426. Also some of the ideas presented here are based on C.H. Edwards's notes 'Hauptvermutung Seminar Notes', University of Georgia (1966), 1–52, and Stallings's article 'On Infinite Processes Leading to Differentiability in the Complement of a Point', *Differential and Combinatorial Topology* (Edited by Stewart S. Cairns), Princeton University Press (1965), 245–254.

Let A be a ring with unit element and $GL(n,A)$ = group of all $n \times n$ invertible matrices over A. We define a homomorphism of $GL(n,A)$ into $GL(n+1,A)$ by sending $M \in GL(n,A)$ to $\begin{pmatrix} M & 0 \\ 0 & 1 \end{pmatrix} \in GL(n+1,A)$ and take $GL(A) = \varinjlim GL(n,A)$. That is, we have inclusions $GL(1,A) \subset GL(2,A) \subset \ldots$. The union $GL(A)$ is called the infinite general linear group.

A matrix is called elementary if it coincides with the identity matrix except for one off-diagonal element. Let $E(A)$ denote the subgroup of $GL(A)$ generated by elementary matrices. $E(A)$ is just the commutator subgroup of $GL(A)$. That is, let aE_{ij} denote the matrix with entry a in the (i,j)th place and zeros elsewhere. The identity $(I+aE_{ij})(I+E_{jk})(I-aE_{ij})(I-E_{jk}) = I+aE_{ik}$, for $i \neq j \neq k \neq i$, shows that each elementary matrix in $GL(n,A)$ is a commutator, providing that $n \geqslant 3$. Conversely the following three identities show that each commutator $XYX^{-1}Y^{-1}$ in $GL(n,A)$ can be expressed as a product of elementary matrices within the larger group $GL(2n,A)$.

$$(1) \quad \begin{pmatrix} XYX^{-1}Y^{-1} & 0 \\ 0 & I \end{pmatrix} = \begin{pmatrix} X & 0 \\ 0 & X^{-1} \end{pmatrix} \begin{pmatrix} Y & 0 \\ 0 & Y^{-1} \end{pmatrix} \begin{pmatrix} (YX)^{-1} & 0 \\ 0 & YX \end{pmatrix},$$

$$(2) \quad \begin{pmatrix} X & 0 \\ 0 & X^{-1} \end{pmatrix} = \begin{pmatrix} I & X \\ 0 & I \end{pmatrix} \begin{pmatrix} I & 0 \\ I\text{-}X^{-1} & I \end{pmatrix} \begin{pmatrix} I & -I \\ 0 & I \end{pmatrix} \begin{pmatrix} I & 0 \\ I\text{-}X & I \end{pmatrix},$$

$$(3) \quad \begin{pmatrix} I & X \\ 0 & I \end{pmatrix} = \prod_{i=1}^{n} \prod_{j=n+1}^{2n} (I + x_{ij} E_{ij}).$$

It follows that $E(A)$ is a normal subgroup of $GL(A)$ with commutative quotient group. The quotient $GL(A)/E(A) = K_1A$ will be called the Whitehead group. We will usually think of K_1A as an additive group. We note that left-multiplication of a matrix M by an elementary matrix corresponds to an elementary row operation. Thus adding multiples of rows by elements of A to rows of a given matrix M gives an equivalent matrix in K_1A. Also K_1 is a covariant functor: that is, any ring homomorphism $A \longrightarrow A'$ gives rise to a group homomorphism $K_1A \longrightarrow K_1A'$.

If the ring A is commutative, then taking determinants of representatives gives a homomorphism of K_1A onto the multiplicative group $U(A)$ of units of A. If A is such that a matrix $M \in GL(A)$ can be reduced to the identity matrix by elementary row operations if and only if $\det M = 1$, then this homomorphism $K_1A \longrightarrow U(A)$ is an isomorphism. For instance, if A is commutative and satisfies a Euclidean algorithm, then $K_1A \cong$ the multiplicative group of units of A. Thus if Z is the integers, then $K_1Z \cong Z_2$ and if F is a field, $K_1F \cong$ multiplicative group of F, F-$\{0\}$.

Given $u \in U(A)$, denote by $[u] \in K_1A$ the element corresponding to the 1×1 matrix $(u) \in GL(1, A) \subset GL(A)$. From the identities

$[uv] = [u] + [v]$ and $[1] = 0$ it follows that $[-1] \in K_1 A$ is either zero or is of order 2. The quotient $K_1 A / \{0, [-1]\} = \overline{K}_1 A$ is called the reduced Whitehead group of A.

Thus for the integers Z the group $\overline{K}_1 Z$ is zero. For the real numbers R, $\overline{K}_1 R$ is isomorphic to the multiplicative group R^+ of positive reals. A specific isomorphism is given by the correspondence $(a_{ij}) \longrightarrow |\det(a_{ij})|$. The advantage of passing to this quotient is that two matrices which differ only by a permutation of rows represent the same element of $\overline{K}_1 A$.

We now assume that A has the property that rank of a finitely generated free module over A is well defined. Then if F is a free A-module with basis $b = (b_1, b_2, \ldots, b_k)$ and basis $c = (c_1, c_2, \ldots, c_k)$, then the class of matrix $M = (a_{ij})$, defined by $c_i = \sum_{j=1}^{k} a_{ij} b_j$, in $\overline{K}_1 A$ is denoted $[c/b]$. That is, $[c/b] \in \overline{K}_1 A$ is the class of the matrix changing b to c. We write $c \sim b$ if $[c/b] = 0$. The identities $[d/c] + [c/b] = [d/b]$, and $[b/b] = 0$ show that this is an equivalence relation.

Given $0 \longrightarrow E \longrightarrow F \longrightarrow G \longrightarrow 0$ an exact sequence of free finitely generated A-modules and given a basis e for E and a basis g for G, then we can construct a basis eg for F via any splitting. That is, if $e = (e_1, \ldots, e_k)$ and $g = (g_1, \ldots, g_\ell)$, lift each $g_i \in G$ to an element $g_i' \in F$. Then $eg' = (e_1, \ldots, e_k, g_1', \ldots, g_\ell')$. The claim is that the equivalence class of eg' is well defined and hence we can write this as eg. That is if $\overline{g} = (\overline{g}_1, \ldots, \overline{g}_\ell)$ is a different lifting, then since $g_i - \overline{g}_i \in E$, we get

$$\begin{pmatrix} e_1 \\ \cdot \\ \cdot \\ \cdot \\ e_k \\ \bar{g}_1 \\ \cdot \\ \cdot \\ \cdot \\ \bar{g}_\ell \end{pmatrix} = \begin{pmatrix} I & 0 \\ & \\ X & I \end{pmatrix} \begin{pmatrix} e_1 \\ \cdot \\ \cdot \\ \cdot \\ e_k \\ g_1' \\ \cdot \\ \cdot \\ \cdot \\ g_\ell' \end{pmatrix}$$

Since the matrix carrying eg' to $e\bar{g}$ is a commutator, $eg' \sim e\bar{g}$. Also we note that $\left[\frac{eg}{\overline{eg}}\right] = \left[\frac{e}{\bar{e}}\right] + \left[\frac{g}{\bar{g}}\right]$. That is, if $e = M\bar{e}$ and $g = N\bar{g}$, then

$$\begin{pmatrix} e \\ g \end{pmatrix} = \begin{pmatrix} M & 0 \\ 0 & I \end{pmatrix} \begin{pmatrix} I & 0 \\ 0 & N \end{pmatrix} \begin{pmatrix} \bar{e} \\ \bar{g} \end{pmatrix}.$$

Now suppose $\mathcal{C}: C_n \xrightarrow{\partial} C_{n-1} \xrightarrow{\partial} \cdots \xrightarrow{\partial} C_1 \xrightarrow{\partial} C_0 \xrightarrow{\partial} 0$ is a free finitely generated chain complex over A with a fixed basis c_i for C_i and $H_i(\mathcal{C})$ is free with a fixed basis h_i, then the torsion $\tau(\mathcal{C}) \in \bar{K}_1 A$ can be defined. For our discussion here, let us assume that for every i, B_i is free (hence Z_i is free by $0 \longrightarrow B_i \longrightarrow Z_i \longrightarrow H_i(\mathcal{C}) \longrightarrow 0$). Choose a basis b_i for B_i (in general B_i will be stably free and $\tau(\mathcal{C})$ still can be defined). Then considering $0 \longrightarrow B_i \longrightarrow Z_i \longrightarrow H_i(\mathcal{C}) \longrightarrow 0$, using bases b_i and h_i we get a basis $b_i h_i$ for Z_i. Then by considering $0 \longrightarrow Z_i \longrightarrow C_i \longrightarrow B_{i-1} \longrightarrow 0$ and using $b_i h_i$ for Z_i and b_{i-1} for B_{i-1}, we get a new basis $(b_i h_i) b_{i-1}$ for C_i (one can show

$(b_i h_i) b_{i-1} = b_i (h_i b_{i-1})$ and hence we will denote this as merely $b_i h_i b_{i-1}$). The torsion of $\mathcal{C}$ is defined $\tau(\mathcal{C}) = \Sigma(-1)^i [b_i h_i b_{i-1}/c_i]$. This does not depend on the choice of the b_i, since, choosing different bases $\bar{b}_i$, we have

$$\Sigma(-1)^i [\bar{b}_i h_i \bar{b}_{i-1}/c_i] = \Sigma(-1)^i ([b_i h_i b_{i-1}/c_i] + [\bar{b}_i/b_i] + [\bar{b}_{i-1}/b_{i-1}]),$$

where the last two terms sum up zero. Of course $\tau(\mathcal{C})$ does depend on the c_i and h_i.

Now consider a short exact sequence $0 \longrightarrow \mathcal{C}' \longrightarrow \mathcal{C} \longrightarrow \mathcal{C}'' \longrightarrow 0$, in the category of chain complexes and chain mappings over A. We will assume that the modules C_i', C_i and C_i'' are free with preferred bases c_i, c_i' and c_i'' which are compatible, in the sense that $c_i \sim c_i' c_i''$. If the homology modules $H_*(\mathcal{C}')$, $H_*(\mathcal{C})$, and $H_*(\mathcal{C}'')$ are all zero, then it can be shown that $\tau(\mathcal{C}) = \tau(\mathcal{C}') + \tau(\mathcal{C}'')$. In general if the homology groups of $\mathcal{C}'$, $\mathcal{C}$, and $\mathcal{C}''$ are not zero, but are free with preferred bases, then if one considers the free acyclic complex $\mathcal{H}: H_n' \to H_n \to H_n'' \to H_{n-1}' \to \ldots \to H_0' \to H_0 \to H_0''$ (and hence $\tau(\mathcal{H})$ is defined), then $\tau(\mathcal{C}) = \tau(\mathcal{C}') + \tau(\mathcal{C}'') + \tau(\mathcal{H})$.

Now let π be a multiplicative group $Z\pi$ the integral group ring (i.e. the set of all finite linear combinations $\Sigma n_i \sigma_i$ where $n_i \in Z, \sigma_i \in \pi$). Then clearly π itself is contained in the group of units $U(Z\pi) \subset GL(1, Z\pi) \subset GL(Z\pi)$. Hence there are natural homomorphisms $\pi \longrightarrow K_1(Z\pi) \longrightarrow \bar{K}_1(Z\pi)$. The cokernel $\bar{K}_1(Z\pi)/\text{image}(\pi)$ is called the Whitehead group $\mathrm{Wh}(\pi)$. This construction can be described by the exact sequence

$$0 \longrightarrow \pi/[\pi,\pi] \longrightarrow \bar{K}_1(Z\pi) \longrightarrow \mathrm{Wh}(\pi) \longrightarrow 0.$$

Wh is a functor from the category of groups to abelian groups. Essentially, in obtaining $\mathrm{Wh}(\pi)$, we consider invertible $n \times n$

matrices (for all n) over $Z\pi$ and make any two such matrices equivalent by:

(i) enlarging by adding 1's along the diagonal (in going to $GL(Z\pi)$),

(ii) adding multiples of rows by elements of $Z\pi$ to rows (in going to $K_1(Z\pi)$),

(iii) interchanging rows (in going to $\overline{K}_1(Z\pi)$),

(iv) multiplying rows by elements of π (in going to $\mathrm{Wh}(\pi)$).

Also $\mathrm{Wh}(\pi)$ is a covariant functor of π. In other words, any homomorphism $f: \pi_1 \to \pi_2$ induces a homomorphism $f_*: \mathrm{Wh}(\pi_1) \longrightarrow \mathrm{Wh}(\pi_2)$. Of particular importance is the following: if $f: \pi \to \pi$ is an inner automorphism, then $f_*: \mathrm{Wh}(\pi) \to \mathrm{Wh}(\pi)$ is the identity. This is clear for let $f(\sigma) = \phi\, \sigma\, \phi^{-1}$ for each $\sigma \in \pi$. The corresponding automorphism of $GL(n, Z\pi)$ is given by

$$(a_{ij}) \longrightarrow \begin{pmatrix} \phi & & & \\ & \cdot & & \\ & & \cdot & \\ & & & \cdot \\ & & & & \phi \end{pmatrix} (a_{ij}) \begin{pmatrix} \phi & & & \\ & \cdot & & \\ & & \cdot & \\ & & & \cdot \\ & & & & \phi \end{pmatrix}^{-1}.$$

Passing to the abelian group $K_1 Z\pi$, or to $\mathrm{Wh}(\pi)$, we therefore obtain the identity automorphism. This remark is of importance when we consider the Whitehead group of a space X, because $\pi_1(X)$ is unique only to within inner automorphisms, due to the arbitrary choice of base point.

It is known that $\mathrm{Wh}(\pi) = 0$ if π is free abelian, free, or is finite of order at most 4. However, if T_5 is the cyclic group of order 5 with generator t, then $\mathrm{Wh}(T_5)$ is infinite cyclic; and the unit $u = (t + t^{-1}-1) \in ZT_5$ represents a generator. (The identity $(t + t^{-1}-1)(t^2 + t^{-2}-1) = 1$ shows that u is a unit).

The relation of Whitehead torsion to geometrical combinatorial topology comes into play by defining the torsion for a CW-complex pair (X,Y), where $|Y|$ is a deformation retract of $|X|$. (Those unfamiliar with CW-complexes may consider the following discussion in terms of simplicial complexes, (K,L) with $|L|$ a deformation retract of $|K|$, as this is where it will be applied).

First we recall some facts from algebraic topology. For any pair (X,Y) of CW-complexes, the associated chain complex $\mathcal{C}(X,Y)$ is defined by setting $C_p(X,Y) = H_p(|X^p \cup Y|, |X^{p-1} \cup Y|)$, where H denotes singular homology with integer coefficients, and where $|X^p|$ denotes the underlying topological space of the p-skeleton of X. This pth chain group is free abelian with one generator for each p-cell of X-Y (that is, let S denote a discrete set which consists of one point s_e from each p-cell of X-Y. Then $|X^{p-1} \cup Y|$ is a deformation retract of $|X^p - S \cup Y|$. Using the exact sequence of the triple $(|X^p \cup Y|, |X^p - S \cup Y|, |X^{p-1} \cup Y|)$: $\longrightarrow$
$H_i(|X^p - S \cup Y|, |X^{p-1} \cup Y|) \longrightarrow H_i(|X^p \cup Y|, |X^{p-1} \cup Y|) \longrightarrow$
$H_i(|X^p \cup Y|, |X^p - S \cup Y|) \longrightarrow H_{i-1}(|X^p - S \cup Y|, |X^{p-1} \cup Y|) \longrightarrow$,
since $H_i(|X^p - S \cup Y|, |X^{p-1} \cup Y|) = 0$ for all i, the middle two terms are isomorphic. By excision the latter of these is isomorphic to $H_i(\cup e, \cup (e - s_e))$ where $\cup e$ denotes the disjoint union of all the p-cells of X-Y. But the homology of such a disjoint union of open sets is the direct sum of the homology groups $H_i(e, e - s_e) \cong H_i(E^p, E^p - 0)$. Since this last expression is free on one generator for $i = p$ and zero otherwise, the above remark follows). Now the homology group $H_p(\mathcal{C}(X,Y))$ of the chain complex $\mathcal{C}(X,Y)$ is canonically isomorphic to the singular group $H_p(|X|, |Y|)$. (That is, $\partial_{p+1} C_{p+1}(X,Y) \to C_p(X,Y)$ is obtained by using the homology exact sequence of the triple $(|X^{p+1} \cup Y|, |X^p \cup Y|, |X^{p-1} \cup Y|)$. Consider the following commutative diagram:

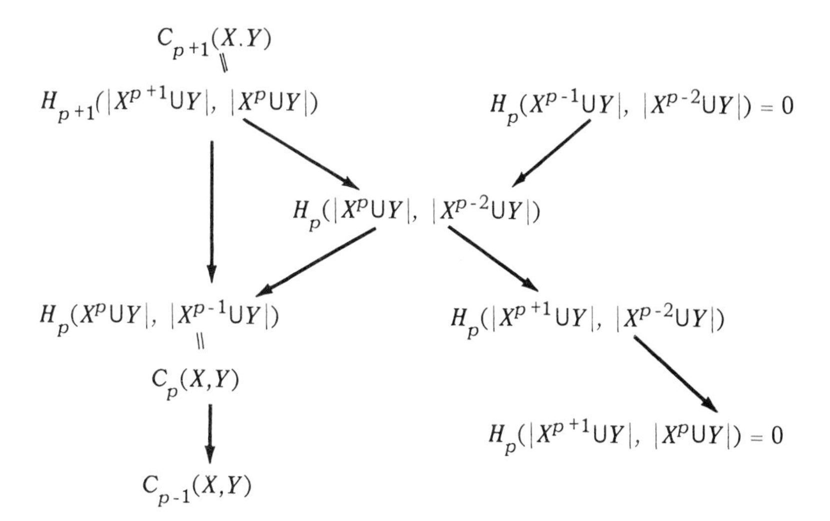

It follows from the diagram that

$Z_p(X,Y) \cong H_p(|X^p \cup Y|, |X^{p-2} \cup Y|)$ and
$Z_p(X,Y)/B_p(X,Y) \cong H_p(|X^{p+1} \cup Y|, |X^{p-2} \cup Y|)$.

Using the homology sequence

$$\to H_p(|X^{p-2} \cup Y|, |Y|) = 0 \to H_p(|X^{p+1} \cup Y|, |Y|) \longrightarrow H_p(|X^{p+1} \cup Y|, |X^{p-2} \cup Y|) \to H_{p-1}(|X^{p-2} \cup Y|, |Y|) = 0 \to$$

we have that $H_p(|X^{p+1} \cup Y|, |X^{p-2} \cup Y|) = H_p(|X^{p+1} \cup Y|, |Y|)$ which is isomorphic to $H_p(|X|, |Y|)$. Thus $H_p(\mathcal{C}(X,Y)) \cong H_p(|X|, |Y|)$.)

Now suppose (K,L) is a finite CW-complex pair where $|L|$ is a deformation retract of $|K|$ and K is connected. Let $p: \hat{K} \longrightarrow K$ be a universal cellular covering map and $(\hat{K},\hat{L})$ ($\hat{L} = p^{-1}(L)$) be the CW-complex pair with respect to which p is cellular (i.e. $\hat{K}$ is a connected simply connected CW-complex and p is a cellular locally trivial fiber map with discrete fiber). Using the fact that, given points $q \in K$, $\hat{q} \in p^{-1}(q)$ and a path $f\colon [0,1] \longrightarrow K$ starting at q, there is a unique path $\hat{f}\colon [0,1] \longrightarrow \hat{K}$ starting at $\hat{q}$ and covering f(i.e., $f = p \circ \hat{f}$), we can easily lift a deformation retraction of K onto L to a deformation retraction of $\hat{K}$ onto $\hat{L}$. Also it follows that the relative homology groups $H_p(|\hat{K}|, |\hat{L}|)$ are trivial.

Now consider the associated chain complex $\mathcal{C}(\hat{K},\hat{L})$ with integer coefficients. In general the chain group $C_p(\hat{K},\hat{L})$ is infinitely generated as an abelian group (being the group of all finite integral linear combinations of oriented p-cells of $(\hat{K}-\hat{L})$. However we can make $C_p(\hat{K},\hat{L})$ into a finitely generated module over the integral group ring $Z\pi$ (where $\pi = \pi_1(K)$) in the following way. Identifying π with the group of covering transformations, each $\sigma \in \pi$ determines a cellular map $\sigma\colon (\hat{K},\hat{L}) \longrightarrow (\hat{K},\hat{L})$ permuting those cells of $\hat{K}$ lying above a given cell of K. Hence σ induces a chain map $\sigma_{\#}\colon \mathcal{C}(\hat{K},\hat{L}) \longrightarrow \mathcal{C}(\hat{K},\hat{L})$. Also given $\alpha = \Sigma n_i\sigma_i \in Z\pi$, we have a chain map $\alpha_{\#}\colon \mathcal{C}(\hat{K},\hat{L}) \longrightarrow \mathcal{C}(\hat{K},\hat{L})$ defined by $\alpha_{\#} = \Sigma n_i\sigma_{i\#}$. This action of $Z\pi$ on $\mathcal{C}(\hat{K},\hat{L})$ clearly makes each chain group $C_p(\hat{K},\hat{L})$ into a $Z\pi$-module which is free with one generator for each p-cell of $K-L$. Since K is finite, it follows that $\mathcal{C}(\hat{K},\hat{L})$ is finitely generated over $Z\pi$.

Thus we now have a free chain complex

$$\mathcal{C}(\hat{K},\hat{L})\colon C_n(\hat{K},\hat{L}) \longrightarrow C_{n-1}(\hat{K},\hat{L}) \longrightarrow \cdots \longrightarrow C_0(\hat{K},\hat{L}) \longrightarrow 0$$

of $Z\pi$-modules, which is exact because $H_i(|\hat{K}|, |\hat{L}|) = 0$. Hence if we were given a fixed basis c_p for each module $C_p(\hat{K},\hat{L})$, then the torsion $\tau(\mathcal{C}(\hat{K},\hat{L})) \in \bar{K}_1(Z\pi)$ would be defined as earlier. Now the geometry of the situation determines a class of preferred bases as follows. Let $e_1, \ldots, e_k$ be the oriented p-cells of $K-L$. For each e_i choose a representative (oriented) p-cell $\hat{e}_i$ of $\hat{K}$ lying over e_i. Then $\hat{e}_i$ determines a basis element of $C_p(\hat{K},\hat{L})$ which we also denote by $\hat{e}_i$. Then $c_p = (\hat{e}_1, \ldots, \hat{e}_k)$ is the preferred $Z\pi$-basis for $C_p(\hat{K},\hat{L})$.

Using these bases, the torsion $\tau(\mathcal{C}(\hat{K},\hat{L}))$ is apparently defined as an element of $\bar{K}_1(Z\pi)$. However, we have made an arbitrary choice of the representative cells e_i. To eliminate the resulting indeterminacy it is necessary to pass to the quotient group $\mathrm{Wh}(\pi)$ and we obtain $\tau(\mathcal{C}(\hat{K},\hat{L}))$ as a class in $\mathrm{Wh}(\pi)$. Note if e_i is replaced by a different representative cell $\pm\sigma_{\#}\hat{e}_i$, then a straightforward verification shows that $\tau(\mathcal{C}(\hat{K},\hat{L}))$ is replaced by $\tau(\mathcal{C}(\hat{K},\hat{L})) - (-1)^p[\sigma] \in \bar{K}_1(Z\pi)$. Thus the difference $\pm[\sigma]$ belongs to image (π), and is annihilated when we pass to the quotient group $\mathrm{Wh}(\pi)$. The image $\tau(\mathcal{C}(\hat{K},\hat{L}))$ in the quotient group $\mathrm{Wh}(\pi)$ is called the Whitehead torsion $\tau(K,L)$ of the CW-complex pair (K,L). We note in making use of $\mathrm{Wh}(\pi) = \mathrm{Wh}(\pi_1(K))$ we never need to worry about base points, since any inner automorphism will induce the identity automorphism of $\mathrm{Wh}(\pi)$.

We will now state some results making use of this torsion.

LEMMA I.20. *If (K,L) is a CW-complex pair as above and each component of $|K| - |L|$ is simply connected, then $\tau(K,L) = 0$.*

Proof. Suppose $|K| - |L|$ has a single component G. If $\hat{G}$ is a component of $|\hat{K}| - |\hat{L}|$, then the simple connectivity of G and the covering homotopy property of the projection p imply that $p|\hat{G}$ is a homomorphism onto G.

If for each k and each cell e^k of $K - L$ we choose the representative cell $\hat{e}^k$ as the unique cell in $\hat{G}$ covering e^k, then the boundary $\partial\hat{e}^k$ is a linear combination with integer coefficients of representative $(k\text{-}1)$-cells. Thus in computing $\tau(\mathcal{C}(\hat{K},\hat{L}))$ we work only with the subring $Z \subset Z\pi$ and it follows that $\tau(\mathcal{C}(\hat{K},\hat{L})) \in \bar{K}_1 Z = 0 \subset \bar{K}_1(Z\pi)$, so $\tau(K,L) = 0$.

If $K\text{-}L$ has several components the proof is essentially the same. It is only necessary to choose a representative component

$\hat{G}_i$ lying over each component G_i of K-L, and to choose representative cells $\hat{e} \subset \hat{G}_i$ as before.

LEMMA I.21. *Given $M \subset L \subset K$ CW-complexes such that $|L|$ and $|M|$ are both deformation retracts of $|K|$ (hence $|M|$ is a deformation retract of $|L|$, then $\tau(K,M) = \tau(K,L) + i_* \tau(L,M)$, where* $i: \pi_1(L) \cong \pi_1(K)$.

The proof here goes back to the algebraic result mentioned after the definition of τ for chain complexes by considering $0 \longrightarrow \mathcal{C}(\hat{L},\hat{M}) \longrightarrow \mathcal{C}(\hat{K},\hat{M}) \longrightarrow \mathcal{C}(\hat{K},\hat{L}) \longrightarrow 0$.

LEMMA I.22. *$K \searrow L$, then $\tau(K,L) = 0$.*

This follows from the above two lemmas by factoring $K \searrow L$ into a sequence $K \searrow K_1 \searrow K_2 \searrow \ldots \searrow K_p \searrow L$ and hence $|K_i| - |K_{i+1}|$ is simply connected.

DEFINITION. A CW pair (M,N) is a subdivision of a CW pair (K,L) if $(|M|, |N|) = (|K|, |L|)$ and every open cell of M lies in some open cell of K.

PROPOSITION I.23. *If (K,L) is a finite CW pair with $|L|$ a deformation retract of $|K|$ and if (M,N) is a subdivision of (K,L), then $\tau(K,L) = \tau(M,N)$.*

The proof here follows from a complicated algebraic subdivision theorem that would have been proven if the algebraic machinery had been developed completely. We note that if $f: (X,Y) \longrightarrow (X_1,Y_1)$ is a PL homomorphism of compact connected polyhedral pairs, then there exist simplicial subdivisions (K,L) and (K_1,L_1) with respect to which f is a simplicial isomorphism. It therefore follows from Proposition I.23 that $f_* \tau(X,Y) = \tau(X_1,Y_1)$. The question of the topological invariance of Whitehead torsion – of whether $f_* \tau(X,Y) = \tau(X_1,Y_1)$ for an arbitrary homomorphism f – remains a fundamental unsolved problem.

Before we go back to obtaining the main results of this section, namely a proof of the weak form of the h-cobordism theorem and giving counterexample to the Hauptvermutung (which made all this discussion necessary), since we have most of the necessary concepts defined, we will state some additional results which give some information about the relation of two simplicial complexes in terms of expansions and collapsings. Recall we have already noted that if $K \searrow L$ then $\tau(K,L) = 0$.

Let L and K be finite simplicial complexes and $f: L \longrightarrow K$ a pwl map, then $M_f = L \times I \cup_{f \times 1} K$ will be called the mapping cylinder of f (i.e., $(l,1)$ is identified with $f(l) \in K$) and M_f will be given a 'natural' triangulation as a simplicial complex. If f is a homotopy equivalence, then $L \times \{0\}$ (which we will denote as merely L) is a deformation retract of M_f and we define $\tau(f) \in \mathrm{Wh}(\pi_1 K)$ which corresponds to $\tau(M_f,L) \in \mathrm{Wh}(\pi_1(M_f))$ under the natural isomorphism $\mathrm{Wh}(\pi_1(M_f)) \cong \mathrm{Wh}(\pi_1(K))$. We say f is a simple homotopy equivalence if $\tau(f) = 0$. Clearly K is a deformation retract of M_f and we have the following result.

LEMMA I.24. *The torsion $\tau(M_f,K)$ is zero.*

Proof. Since there is a subdivision of (M_f,K) such that $M_f \searrow K$, Lemma I.22 and Proposition I.23 give the desired results. We will also give a proof that is valid for L and K CW-complexes and M_f the CW mapping cylinder of a cellular map f. Let $f(p): L^p \longrightarrow K$ denote the restriction of f to the p-skeleton of L, so that $K = M_{f(-1)} \subset M_{f(0)} \subset \ldots \subset M_{f(n)} = M_f$. Then $\tau(M_f,K) = \Sigma_p \tau(M_{f(p)}, M_{f(p-1)})$ by repeated applications of Lemma I.21. But each term on the right is zero by Lemma I.20.

LEMMA I.25. *If L is a deformation retract of K and $f: L \longrightarrow K$ is an inclusion, then $\tau(f) = \tau(K,L)$.*

Proof. $\tau(f) = \tau(M_f, L \times 0) = \tau(M_f, L \times I) + \tau(L \times I, L \times 0) = \tau(M_f, L \times I)$ since $L \times I \searrow L \times 0$. Now $\tau(M_f, L \times 1) = \tau(M_f, L \times I) + \tau(L \times I, L \times 1)$ and hence $\tau(M_f, L \times I) = \tau(M_f, L \times 1) = \tau(M_f, K) + \tau(K, L \times 1)$. But $\tau(M_f, K) = 0$ by Lemma I.24 and the result follows.

LEMMA I.26. *If f_0 and f_1 are homotopic, then $\tau(f_0) = \tau(f_1)$.*

LEMMA I.27. *If $f: X \longrightarrow Y$ and $g: Y \longrightarrow Z$ are piecewise linear homotopy equivalences, then*

$$\tau(g \circ f) = \tau(g) + g_* \tau(f).$$

(Here $g: Y \longrightarrow Z$ induces g_*: $\mathrm{Wh}(\pi_1(Y)) \longrightarrow \mathrm{Wh}(\pi_1(Z))$.)

It will be convenient to prove a more general statement which implies both I.26 and I.27. Let $h: X \longrightarrow Z$ be any pwl mapping which is pwl homotopic to $g \circ f$.

Proof that $\tau(h) = \tau(g) + g_* \tau(f)$: Choose a pwl homotopy $X \times [0,1] \longrightarrow Z$ between h and $g \circ f$. This homotopy can be pieced together with the mapping $g: Y \longrightarrow Z$ to yield a pwl mapping $H: M_f \longrightarrow Z$ where $H | X \times \{0\} = h$ and $H | Y = g$.

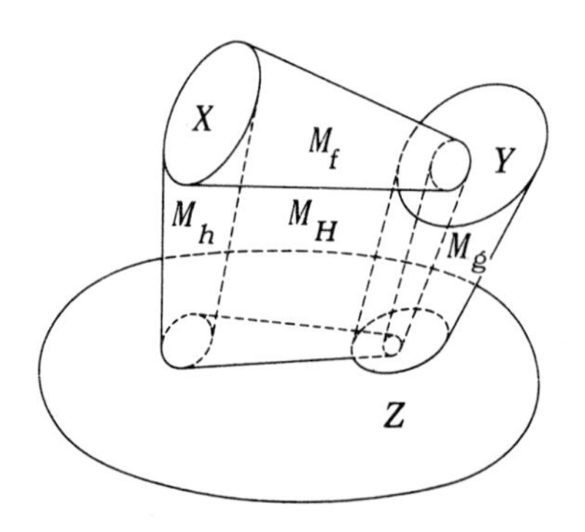

Note that the mapping cylinder M_H contains the mapping cylinders of f, g and h as subcomplexes. In fact we have the following diagram of inclusion maps.

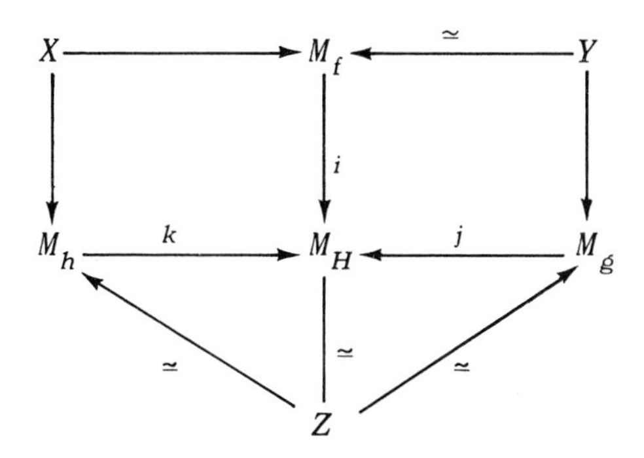

By Lemma I.24, $\tau(M_f, Y) = 0$, $\tau(M_g, Z) = 0$, $\tau(M_H, Z) = 0$, and $\tau(M_h, Z) = 0$. We have indicated this by inserting the symbol $\simeq$ (for simple homotopy equivalence) on the appropriate arrows.
Now by Lemma I.21 $\tau(M_H, Z) = \tau(M_H, M_g) + j_* \tau(M_g, Z)$ and $\tau(M_H, Z) = \tau(M_H, M_h) = k_* \tau(M_h, Z)$. Thus $\tau(M_H, M_g) = 0 = \tau(M_H, M_h)$. That is j and k are simple homotopy equivalences. By the right hand square we see that

$$\underbrace{\tau(M_H, M_f)}_{} + i_* \underset{\substack{\| \\ 0}}{\tau(M_f, Y)} = \tau(M_H, Y) = \underset{\substack{\| \\ 0}}{\tau(M_H, M_g)} + j_* \tau(M_g, Y).$$

Similarly from the left hand square:

$$0 + k_* \tau(M_h, X) = \tau(M_H, M_f) = i_* \tau(M_f, X).$$

Hence

$$k_* \tau(M_h, X) = j_* \tau(M_g, Y) + i_* \tau(M_f, X).$$

Now applying the isomorphism $\mathrm{Wh}(\pi_1(M_H)) \xrightarrow{\cong} \mathrm{Wh}(\pi_1(Z))$ to the above, we obtain the required equation

$$\tau(h) = \tau(g) + g_* \tau(f).$$

This completes the proof of I.26 and I.27.

Note it follows from I.26 that the torsion can be defined even for a homotopy equivalence $X \longrightarrow Y$ which is not pwl, for any map between simplicial complexes is homotopic to a simplicial map.

DEFINITION. If X and Y are finite simplicial complexes, then we say $X \equiv Y$ (X and Y have the same nucleus), if there exists a sequence of complexes $X = X_1, X_2, \ldots, X_k = Y$ such that either $X_i \searrow X_{i+1}$, $X_i \nearrow X_{i+1}$, or $X_i \approx X_{i+1}$. $X \underset{n}{\equiv} Y$ means there exists a sequence as above with $\dim X_i \leqslant n$ for all i.

THEOREM I.28. *If $X \equiv Y$, then there exists a simple homotopy equivalence $f\colon X \longrightarrow Y$.*

Proof. Since $X \equiv Y$, there exists a sequence $X = X_1, X_2, \ldots, X_{k-1}, X_k = Y$, where either $X_i \searrow X_{i+1}$, $X_i \nearrow X_{i+1}$, or $X_i \approx X_{i+1}$. Each of these induce a homotopy equivalence $f_i\colon X_i \longrightarrow X_{i+1}$ where either f_i is a retraction of X_i onto X_{i+1}, f_i is an inclusion of X_i into X_{i+1} with X_i a deformation retract of X_{i+1}, or f_i is a pwl homomorphism. If $f_i\colon X_i \longrightarrow X_{i+1}$ is a retraction, then $\tau(f_i) = 0$; for if $j\colon X_{i+1} \subset X_i$, then $f_i \circ j = 1_{X_{i+1}}$ and $0 = \tau(1) = \tau(f_i \circ j) = \tau(j) + \tau(f_i) = 0 + \tau(f_i)$. $\tau(j) = 0$, since $\tau(j) = \tau(X_i, X_{i+1})$ and $X_i \searrow X_{i+1}$. Similarly, if $f_i\colon X_i \subset X_{i+1}$ is an inclusion, then $\tau(f_i) = \tau(X_{i+1}, X_i) = 0$ since $X_{i+1} \searrow X_i$. If $f_i\colon X_i \longrightarrow X_{i+1}$ is a pwl homomorphism, $\tau(f_i) = 0$ by the subdivision theorem. Finally taking $f = f_{k-1} \circ \cdots \circ f_2 \circ f_1$, it follows from I.27 that $\tau(f) = 0$.

THEOREM I.29. *Suppose X and Y have the same simple homotopy type (i.e., have homotopy equivalence $f\colon X \longrightarrow Y$ where $\tau(f) = 0$), then $X \underset{n+1}{\equiv} Y$ where $n = \max\,[(\dim X) + 1,\ \dim Y, 3]$.*

Before we prove Theorem I.29 we will first give a definition and obtain a lemma. Suppose M is a finite simplicial complex, $\dim M = n$, and M^k denotes the k-skeleton of M. By a simplicial cellular decomposition of M we mean there exists for each $k = 0, 1, \ldots, n$ a finite number of pairs of maps f_j^k, g_j^k, $j = 1, \ldots m_k$ of a k-simplex Δ^k into $|M|$ so that:

(1) For each j there is a subdivision α_j of Δ^k so that $g_j^k\colon \alpha_j \Delta^k \longrightarrow M$ is simplicial;

(2) for each j, $f_j^k\colon \Delta^k \longrightarrow |M|$ is a map with $f_j^k(\Delta^k) = |g_j^k(\Delta^k)|$, $f_j^k \mid \text{int}\,\Delta^k$ is a homomorphism, $f_i^k \mid_{\text{Bd}\Delta^k} = g_j^k \mid_{\text{Bd}\Delta^k}$, and $f_j^k(\text{Bd}\Delta^k) \subset \bigcup_{i=0}^{k-1} (\bigcup_{j=1}^{m_i} f_j^i(\text{int}\,\Delta^i))$;

(3) for $j \neq i$, $f_j^k(\text{int}\,\Delta^k) \cap f_i^k(\text{int}\,\Delta^k) = \emptyset$; and

(4) $\bigcup_{k=0}^{n} (\bigcup_{j=1}^{m_k} f_j^k(\text{int}\,\Delta^k)) = |M|$.

We will denote such a simplicial cellular decomposition by $C(M)$. The k-cells of $C(M)$ will be the $f_i^k(\text{int}\,\Delta^k)$, $j = 1, \ldots, m_k$. The k-skeleton, $C(M)^k$, will be the union of all i-cells of $C(M)$ for $i \leqslant k$. We note

$$C(M)^k = \left(\bigcup_{i=0}^{k} \bigcup_{j=1}^{m_i} f_j^i(\text{int}\,\Delta^i)\right) = \left(\bigcup_{i=0}^{k} \bigcup_{j=1}^{m_i} f_j^i(\Delta^i)\right)$$

$$= \left(\bigcup_{i=0}^{k} \bigcup_{j=1}^{m_i} |g_j^i(\alpha_j \Delta^i)|\right)$$

and hence $C(M)^k$ is covered by a subcomplex of M contained in M^k. Thus $C(M)$ gives a CW-complex structure on M 'compatible' with M as a simplicial complex.

LEMMA FOR I.29. *Suppose that* $X \subset M$ *are simplicial complexes where* $|X|$ *is a deformation retract of* $|M|$, dim $M = n$, and $\tau(M,X) = 0$. *If* $C(M)$ *is a simplicial cellular decomposition of* M *inducing a simplicial cellular decomposition* $C(X)$ *of* X *so that for some* $k \leqslant n\text{-}3$, $C(M)^k \subset |X|$, *then there exist complexes* $X' \subset M'$ *and a simplicial cellular decomposition* $C(M')$ *of* M' *inducing a simplicial cellular decomposition* $C(X')$ *of* X' *so that* $C(M')^{k+1} \subset |X'|$, $X' \underset{n}{\equiv} X, M' \underset{n}{\equiv} M$, $|X'|$ *is a deformation retract of* $|M'|$ *and* $\tau(M',X') = 0$.

Proof. Suppose $f_j^{k+1}(\Delta^{k+1})$ is not contained in $|X|$. Let Δ^{k+2} be a $(k+2)$-simplex with Δ^{k+1} a face of Δ^{k+2}. Since $f_j^{k+1}(\text{Bd}\Delta^{k+1}) \subset |X|$ and $|X|$ is a deformation retract of $|M|$, there exists a pwl map h_j of Δ^{k+2} into M so that $h_j|_{\Delta^{k+1}} = g^{k+1}|_{\Delta^{k+1}}$ and $h_j(\text{Bd}\Delta^{k+2}\text{-int }\Delta^{k+1}) \subset |X|$. Since $C(X) \subset C(M)$ and the closure of each i-cell of $C(M)$ is covered by a simplicial subcomplex of M, we may obtain such an h_j so that $h_j(\Delta^{k+2}) \subset C(M)^{k+2}$ and $h_j(\text{Bd}\Delta^{k+2}) \subset C(M)^{k+1}$. Let M' be $M \cup \Delta^{k+2} \times [0,1]$ with $(x,1)$ identified with $h_j(x)$ subdivided so that M' is a simplicial complex. Let X' be the subcomplex of M' corresponding to $X \cup \Delta^{k+2} \times \{0\} \cup \text{Bd}\Delta^{k+2} \times [0,1]$. Now $X' \searrow$ subdivision of X and $M' \searrow$ subdivision of M, hence $X' \underset{n}{\equiv} X$, $M' \underset{n}{\equiv} M$. Since $\pi_i(X') \cong \pi_i(X) \cong \pi_i(M) \cong \pi_i(M')$ we have $\pi_i(M',X') = 0$ and hence $|X'|$ is a deformation retract of $|M'|$. $\tau(M', X) = \tau(M', X') + \tau(X', X)$ and $\tau(M', X) = \tau(M', M) + \tau(M, X)$. Since $\tau(X', X) = 0 = \tau(M', M)$ we get $\tau(M', X') = \tau(M, X) = 0$. The cellular structure on M' is that given for $|M| \subset |M'|$ (although the simplicial structure on the closure of the i-cells is subdivided now) plus a $(k+3)$-cell corresponding to int $\Delta^{k+2} \times (0,1)$ and a $(k+2)$-cell corresponding to int $(\Delta^{k+2} \times \{0\} \cup \text{Bd}\Delta^{k+2} \times [0,1])$. We note that the latter $(k+2)$-cell lies in $C(X')$ and that also $f_j^{k+1}(\Delta^{k+1}) \subset C(X')$. We now repeat this procedure for each $f_i^{k+1}(\Delta^{k+1})$ that is not contained in $|X|$ and the result follows.

Proof of Theorem I.29. First we may as well assume that $f: X \longrightarrow Y$ is a homotopy equivalence with $\tau(f) = 0$ where dim $X \geqslant 1$. Also if both dim X and dim Y are $\geqslant 1$, then we may suppose dim $X \leqslant$ dim Y. That is, if $f: X \longrightarrow Y$ is a homotopy equivalence, then there is a homotopy equivalence $g: Y \longrightarrow X$ such that $g \circ f \cong 1_X$. But then $\tau(1_X) = \tau(g \circ f) = \tau(g) + g_* \tau(f)$. Hence if $\tau(f) = 0$, then $\tau(g) = 0$. Also we can assume that the dimension of one of X or Y is $\geqslant 1$, otherwise $X \approx Y$ trivially.

Let M be the mapping cylinder of $f: X \longrightarrow Y$. Then we have $X\ (=X \times \{0\}) \subset M$ and $Y \subset M$, $\tau(M,X) = \tau(f) = 0$ and $M \underset{n}{\equiv} Y$ (since $M \searrow Y$) where max (dim $X+1$, dim Y) $= n \geqslant 2$. Later we will want $n \geqslant 3$. If such is not the case we can add 3-simplexes to M trivially so that dim $M = n \geqslant 3$ and $M \underset{n}{\equiv} Y$, where now $n = \max(\dim X + 1, \dim Y, 3)$. We now can make repeated applications of the lemma for I.29 starting with $-1 = k \leqslant n - 3$ (certainly $M^{-1} \subset X$) to obtain complexes X', M' with $X' \subset M'$, $X' \equiv X$, $M' \equiv M$, $|X'|$ is a deformation retract of $|M'|$, $(M', \dot{X}') = 0$, and M' has a simplicial cellular decomposition $C(M')$ inducing $C(X') \subset C(M')$ of X' so that $C(M')^{n-2} \subset |X'|$.

Thus we have CW-complexes $C(X') \subset C(M')$ with $C(M')^{n-2} \subset C(X')$, dim $C(M') = n$, $|C(X')|$ is a deformation retract of $|C(M')|$, and $\tau(C(M'), C(X')) = 0$. The associated chain complex for $(\widehat{C(M')}, \widehat{C(X')})$ then has the form

$$\mathcal{C}(\widehat{C(M')}, \widehat{C(X')}) : H_n(|\widehat{C(M')} \cup \widehat{C(X')}|, |\widehat{C(M')}^{n-1} \cup \widehat{C(X')}|) \xrightarrow{\partial} H_{n-1}(|\widehat{C(M')}^{n-1} \cup \widehat{C(X')}|, |\widehat{C(M')}^{n-1} \cup \widehat{C(X')}|) \longrightarrow 0.$$

Now if $\hat{e}_1'^{n-1}, \ldots, \hat{e}_k'^{n-1}$ is a preferred basis for H_{n-1} and $\hat{e}_1'^{n}, \ldots, \hat{e}_k'^{n}$ is a preferred basis for H_n, then $\partial \hat{e}_i'^{n} = \Sigma_j a_{ij}\ \hat{e}_j'^{n-1}$ and

$\tau(C(M'), C(X')) = 0$ is the class in $\mathrm{Wh}(\pi)$ of (a_{ij}). Since this class is zero this means that (a_{ij}) can be enlarged (by adding 1's along the diagonal – which corresponds geometrically to adding trivial elements to $C(M')$) and then reduced to the identity by the following operations:

(1) interchange two rows,

(2) multiply a row by $\pm$ an element of π, and

(3) add a $Z\pi$-multiple of one row to another.

Since operations of type (1) and (2) are trivial – they correspond to re-ordering the bases or using different representatives of the $Z\pi$-basis elements – we only have to consider operations of type (3). Here it is sufficient just to consider the operation of adding a π-multiple of one row to another since π generates $Z\pi$.

Thus, suppose (a'_{ij}) is a matrix obtained from (a_{ij}) by adding a π-multiple of one row to another. That is, we can suppose that (a'_{ij}) has the form $a'_{i_1 j} = a_{i_1 j} + \sigma a_{i_2 j}$, $\sigma \in \pi = \pi_1(M')$, and $a'_{ij} = a_{ij}$ for $i \neq i_1$. For notational purposes let us suppose $a'_{1j} = a_{1j} + \sigma a_{2j}$ and $a'_{ij} = a_{ij}$ for $i \geqslant 2$.

The claim is there exists a simplicial map $h\colon \Delta^n \longrightarrow |\widehat{C(M')}|$ so that $h(\mathrm{Bd}\Delta^n)$ represents the element $\partial(\hat{e}_1'^n + \sigma\, \hat{e}_2'^n)$ of $H_{n-1}(|\widehat{C(M')}^{n-1} \cup \widehat{C(X')}|, |\widehat{C(M')}^{n-2} \cup \widehat{C(X')}|) = H_{n-1}(|\widehat{C(M')}^{n-1} \cup \widehat{C(X')}|, |\widehat{C(X')}|)$. Since $(\widehat{C(M')}, \widehat{C(X')})$ is the universal cover of $(C(M'), C(X'))$, $|\widehat{C(X')}|$ is a deformation retract of $|\widehat{C(M')}|$ and $\widehat{C(M')}^{n-2} \subset \widehat{C(X')}$, it follows that if we suppose that $n \geqslant 3$ then $\pi_1(|\widehat{C(M')}^{n-1} \cup \widehat{C(X')}|) = 0 = \pi_1(\widehat{C(X')})$ and that $H_q(|\widehat{C(M')}^{n-1} \cup \widehat{C(X')}|, |\widehat{C(X')}|) = 0$ for $q < n-1$. Hence by the relative Hurewicz isomorphism theorem we have the $\pi_q(|\widehat{C(M')}^{n-1} \cup \widehat{C(X')}|, |\widehat{C(X')}|) = 0$ for $q < n-1$ and $\pi_{n-1}(|\widehat{C(M')}^{n-1} \cup \widehat{C(X')}|, |\widehat{C(X')}|) \cong H_{n-1}(|\widehat{C(M')}^{n-1} \cup \widehat{C(X')}|, |\widehat{C(X')}|)$. Thus given any simplicial map

$\hat{h}: \Delta^{n-1} \longrightarrow |\widehat{C(M')}^{n-1} \cup \widehat{C(X')}|$ so that $\hat{h}(\text{Bd}\Delta^{n-1}) \subset |C(X')|$, since $\pi_{n-1}(|\widehat{C(M')} \cup \widehat{C(X')}|, |\widehat{C(X')}|) = 0$, there exists a simplicial map $h: \Delta^n \longrightarrow |\widehat{C(M')}|$ so that $h|_{\Delta^{n-1}} = \hat{h}$ and $h(\text{Bd}\Delta^n\text{-int }\Delta^{n-1}) \subset |\widehat{C(X')}|$. Now h on $\text{Bd}\Delta^n$ and $\hat{h}$ on Δ^{n-1} represent the same element H_{n-1} $(|\widehat{C(M')}^{n-1} \cup \widehat{C(X')}|, |\widehat{C(X')}|)$.

Thus suppose h taking Δ^n into $|C(M')|$ has the property that $h|_{\text{Bd}\Delta^n}$ represents the element $\partial(\hat{e}_1'^n + \sigma\, \hat{e}_2'^n)$. Let $\tilde{h}$ be the simplicial map $p \circ h$ taking Δ^n into M' $(p : \hat{M}' \longrightarrow M'$ is the covering projection). Let M_1'' be the simplicial complex obtained by attaching some subdivision of an n-simplex, Δ^n, to $M' \text{-} e_1'^n$ by the map $\tilde{h}|_{\text{Bd}\Delta^n}$. That is, $M_1'' = (M' \text{-} e_1'^n) \cup_{h|\text{Bd}\Delta^n} \Delta^n$. $C(M_1'')$ has the simplicial cellular decomposition given by $C(M')$. Except now, $e_1'^n$ is replaced by the n-cell $e_1''^n$ which corresponds to the interior of the new n-simplex we have just attached. X_1'' is just X' under the appropriate subdivision.

Now using the basis $\hat{e}_1''^n, \hat{e}_2'^n, \ldots, \hat{e}_k'^n$ for H_n and $\hat{e}_1'^{n-1}$, $\hat{e}_2'^{n-1}, \ldots, \hat{e}_k'^{n-1}$ for H_{n-1} of the associated chain complex $\mathcal{C}(\widehat{C(M_1'')}$, $\widehat{C(X_1'')})$, we have $\partial\, \hat{e}_i'^n = \Sigma_{j=1}^k\, a_{ij}'\, \hat{e}_j'^{n-1}$ and thus we obtain the matrix (a_{ij}'). Since $H_p(\mathcal{C}(\widehat{C(M_1'')}, C(X_1''))) \cong H_p(|\hat{M}_1''|, |\hat{X}_1''|)$ and (a_{ij}') is a non-singular, it follows that $H_p(|\hat{M}_1''|, |\hat{X}_1''|) = 0$ for all p. But then $H_*(|\hat{M}_1''|, |\hat{X}_1''|) \cong \pi_*(|\hat{M}_1''|, |\hat{X}_1''|) \cong \pi_*(|M_1''|, |X_1''|) = 0$ and $|X_1''|$ is a deformation retract of $|M_1''|$.

Suppose g_1' is the simplicial map of a subdivision of Δ^n into M' where $|g_1'(\Delta^n)| = f_1'(\Delta^n)$ and $f_1'(\text{int }\Delta^n)$ gives the cellular n-cell $e_1'^n$ of $C(M')$. Let $\tilde{g}$ be the simplicial map of a subdivision of Δ^n into M_1'', where $|\tilde{g}(\Delta^n)| = \tilde{f}(\Delta^n)$ and $\tilde{f}(\text{int }\Delta^n)$ gives the cellular n-cell $e_1''^n$ of $C(M_1'')$. We also may suppose that $\tilde{g}|_{\text{Bd}\Delta^n} = \tilde{h}|_{\text{Bd}\Delta^n}$. Now there exists a simplicial map $k : \text{Bd}(\Delta^n \times I) \longrightarrow M' \cup M_1''$, where $M' \cap M_1'' = M' \text{-} e_1'^n$, so that $k|_{\Delta^n \times 0} = g_1'$, $k|_{\Delta^n \times 1} = \tilde{g}$ and $k(\text{Bd}\Delta^n \times I) \subset M' \text{-} e_1'^n$ (since $g_1'|_{\text{Bd}\Delta^n}$ is homotopic to $\tilde{h}|_{\text{Bd}\Delta^n}$ in

$M' - e_1'^n$). Let Z be the simplicial complex formed from $(M' \cup M_1'')$ $\cup_k (\Delta^n \times I)$ under an appropriate subdivision. Then there are subdivisions $\tilde{\alpha}$, α' of Z so that $\tilde{\alpha}Z \searrow \tilde{\alpha}M'$ (pushing in from $\tilde{g}(\Delta^n \times 1)$) and so that $\alpha' Z \searrow \alpha' M_1''$ (pushing in from $g'(\Delta^n \times 0)$). This follows by the following more general result.

Assertion. Let N be a combinatorial manifold with non-empty boundary $\dot{N}$ and suppose L is a subcomplex of $\dot{N}$ so that the following is true. (i) We are given a subdivision α of L and a simplicial map g taking αL into a complex K. (ii) We can extend αL to a subdivision αN so that if R is the simplicial neighborhood of $(\alpha L)''$ in $(\alpha N)''$ (that is, R is a regular neighborhood) then for some subdivision $\hat{\alpha}$ of $(\alpha N)''$, $\hat{\alpha}(\alpha N)'' \searrow \hat{\alpha}R$. Then $K \cup_g N$ has a 'natural' triangulation as a simplicial complex – say $\gamma K \cup_g \beta N$ so that g taking $\beta L \longrightarrow \gamma K$ is simplicial and $\gamma K \cup_g \beta N \searrow \gamma K \cup_g (\beta L) = \gamma K$.

Proof of the assertion. Let $\dot{R}$ denote the subcomplex of R such $|\dot{R}|$ is the topological boundary of R in N as a point set (i.e. $|\dot{R}| = |R| \cap \overline{(N\text{-}R)}$).

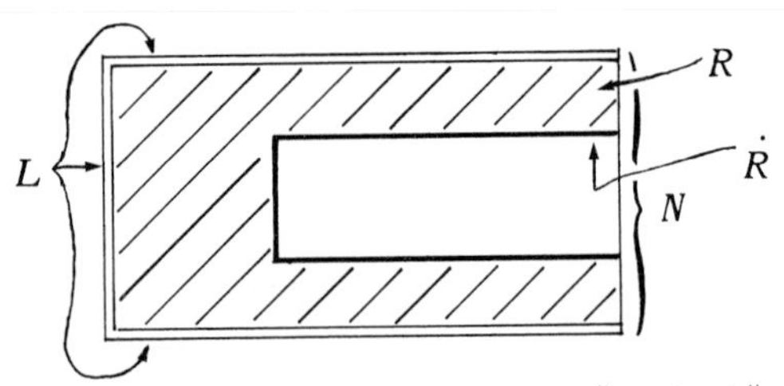

Since R is a regular neighborhood of $(\alpha L)''$ in $(\alpha N)''$ and $|L| \subset \text{int}|R|$ (where interior is taken with respect to $|N|$), there is a pwl map φ: $\hat{\alpha}\dot{R} \longrightarrow \hat{\alpha}\,(\alpha L)''$ so that the mapping cylinder $[(\hat{\alpha}\dot{R}) \times I] \cup_\varphi \hat{\alpha}(\alpha L)'' \approx R$. We then have that $\hat{\alpha}(\alpha N) \approx \text{Cl}(\hat{\alpha}(\alpha N)'' - \hat{\alpha}R) \cup ([\hat{\alpha}\dot{R}) \times I] \cup_\varphi \hat{\alpha}\,(\alpha L)'')$. (Here Cl denotes the complex covering the closure of the given set). Now there is a subdivision $\hat{\gamma}$ of K so that the composition

$\hat{\alpha}\dot{R} \xrightarrow{\varphi} \hat{\alpha}(\alpha L)'' \xrightarrow{g} \hat{\gamma}K$ is simplicial. We then have that $K \cup_g N \approx ([(\hat{\alpha}\dot{R}) \times I] \cup_{g\circ\varphi} \hat{\gamma}K) \cup \mathrm{Cl}\,(\hat{\alpha}(\alpha N)'' - \hat{\alpha}R)$. Since $\hat{\alpha}(\alpha N)'' \searrow \hat{\alpha}R$ and $(\hat{\alpha}R) \times I \cup_{g\circ\varphi} \hat{\gamma}K \searrow \hat{\gamma}K$ we can obtain subdivisions γ of K and β of N so that $g : \beta L \to \gamma K$ is simplicial and $\gamma K \cup_g \beta N \searrow \gamma K$.

Now, it follows easily from the above result, that there exist subdivisions $\tilde{\alpha}$ and α' so that $\tilde{\alpha} Z \searrow \tilde{\alpha} M'$ and $\alpha' Z \searrow \alpha' M_1''$. Hence $M' \underset{n+1}{\equiv} Z \underset{n+1}{\equiv} M_1''$, $X_1'' \underset{n}{\equiv} X''$, $X_1'' \subset M_1''$, $\tau(M_1'', X_1'') = 0$, and the boundary operator ∂ in $\mathcal{C}(\widehat{C(M_1'')}, \widehat{C(X_1'')})$ is given by (α_{ij}'). Now since our original (α_{ij}) is equivalent to the identity matrix in $\mathrm{Wh}(\pi)$, we can repeat the above procedures a finite number of times and obtain complexes $X'' \subset M''$ so that, $\dim M'' = n$, $|X''|$ is a deformation retract of $|M''|$, $\tau(M'', X'') = 0$, $M'' \underset{n+1}{\equiv} M'$, $X'' \underset{n}{\equiv} X'$, M'' has a simplicial cellular decomposition $C(M'')$ with $C(X'') \subset C(M'')$ and $C(M'')^{n-2} \subset C(X'')$, and in considering the associated chain complex $\mathcal{C}(\widehat{C(M'')}, \widehat{C(X'')})$ we have $\partial \hat{e}_i''^{\,n} = \hat{e}_i''^{\,n-1}$ for $i = 1, \ldots, m$. That is, the matrix (α_{ij}'') is now just the identity matrix. Then for each n-cell $e_i''^{\,n}$ of $C(M'') - C(X'')$ we have that $\partial e_i''^{\,n} = e_i''^{\,n-1} \pmod{C(X'')}$, e_i^{n-1} lies in $C(M'') - C(X'')$, and $(\cup_{i=1}^m e_i''^{\,n}) \cup (\cup_{i=1}^m e_i''^{\,n-1}) = C(M'') - C(X'')$.

The claim is that there is a subdivision $(\tilde{M}, \tilde{X})$ of the pair (M'', X'') so that $\tilde{M} \searrow \tilde{X}$. That is, for each $i = 1, \ldots, m$ we may suppose that we have a 4-tuple of maps $(f_i^n, g_i^n, f_i^{n-1}, g_i^{n-1})$ having the following properties.

(i) g_i^n is a simplicial map of an n-simplex Δ^n into M'' and f_i^n is a map of Δ^n into $|M''|$ so that $|g_i^n(\Delta^n)| = f_i^n(\Delta^n)$, $f_i^n|_{\mathrm{int}\,\Delta^n}$ is a homomorphism where $f_i^n(\mathrm{int}\,\Delta^n)$ gives the cellular n-cell $e_i''^{\,n}$ and $f_i^n|_{\mathrm{Bd}\Delta^n} = g_i^n|_{\mathrm{Bd}\Delta^n}$;

(ii) g_i^{n-1} is a simplicial map of $\Delta^{n-1} \subset \Delta^n$ into M'' so that $g_i^n|_{\Delta^{n-1}} = g_i^{n-1}$ and f_i^{n-1} is a map of Δ^{n-1} into $|M''|$ so that $|g_i^{n-1}(\Delta^{n-1})| = f_i^{n-1}(\Delta^{n-1})$, $f_i^{n-1}|_{\mathrm{int}\,\Delta^{n-1}}$ is a homomorphism where

$f_i^{n-1}(\text{int}\ \Delta^{n-1})$ gives the cellular $(n\text{-}1)$-cell $e_i''^{n-1}$ and $f_i^{n-1}|\text{Bd}\ \Delta^{n-1} = g_i^{n-1}|_{\text{Bd}\Delta^{n-1}}$; and

(iii) $g_i^n(\text{Bd}\Delta^n - \text{int}\ \Delta^{n-1}) \subset X''$ (since $\partial e_i''^n = e_i''^{n-1}$ (mod $C(X'')$)). Since $C(M'') - C(X'') = (\cup_{i=1}^m e_i''^n) \cup (\cup_{i=1}^m e_i''^{n-1})$ and the $\{e_i''^{n-1}\}$ are all disjoint, each $e_i''^{n-1}$ is a 'free face' of $e_i''^n$ in $C(M'')$. We now can make use of the above result we used in showing that $M' \underset{n+1}{\equiv} Z \underset{n+1}{\equiv} M_1''$. That is, we can obtain a subdivision α_i of $g_i^n(\Delta^n)$ so that $\alpha_i g^{n-1}(\Delta^{n-1}) \searrow \alpha_i g_i^n(\text{Bd}\Delta^n\text{-int}\ \Delta^{n-1}) \subset \alpha_i X''$ by 'pushing in' from the 'free face' $\alpha_i g_i^{n-1}(\Delta^{n-1})$. But then there is a subdivision β of X'' and subdivisions β_i of $\alpha_i g_i^n(\text{Bd}\Delta^n\text{-int}\Delta^{n-1})$ so that $\beta_i \alpha_i g_i^n(\text{Bd}\Delta^n\text{-int}\ \Delta^{n-1})$ is a subcomplex of $\beta X''$ for each i. By Lemma IV.6 of Volume I, there is a subdivision γ_i of $\alpha_i g_i^n(\Delta^n)$ so that $\beta_i \alpha_i g_i^n(\text{Bd}\Delta^n\text{-int}\ \Delta^{n-1})$ is a subcomplex of $\gamma_i \alpha_i g_i^n(\Delta^n)$ and $\gamma_i \alpha_i g_i^n(\Delta^n) \searrow \beta_i \alpha_i g_i^n(\text{Bd}\Delta^n\text{-int}\ \Delta^{n-1})$. Thus taking $\tilde{X}$ to be $\beta X''$ and $\tilde{M}$ to be $\beta X \cup (\cup_{i=1}^m \gamma_i \alpha_i g_i^n(\Delta^n))$ we get that $\tilde{M} \searrow \tilde{X}$.

Summarizing, we started with two complexes X and Y and we were given a homotopy equivalence $f: X \to Y$ with $\tau(f) = 0$, where we assumed that dim $X \geqslant 1$ and that dim $X \leqslant$ dim Y if both X and Y had dimension $\geqslant 1$. We then obtained the following diagram:

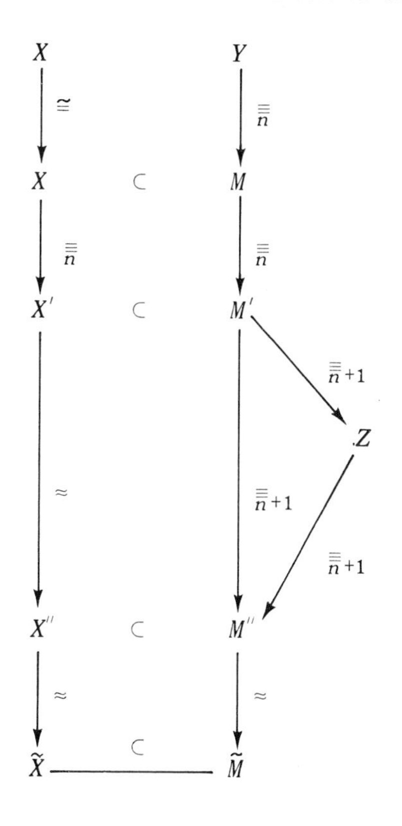

Here M is the mapping cylinder of $f; X \to Y$, enlarged if necessary, so that $n = \dim M = \max(\dim X + 1, \dim Y, 3)$. Here we have $C(M') \subset^{n-2} C(X')$.

Here ∂ in $\mathcal{C}(C(M''), C(X''))$ gives the identity matrix.

Here $\tilde{M} \searrow \tilde{X}$ and this completes the proof of Theorem I.29.

That is, our desired sequence of complexes is obtained by starting with X and just following the above diagram around to Y.

Now we return to the main theme of this section. Suppose $(W; M, M')$ is a PL cobordism. A handlebody presentation of W is a sequence $W^{(-1)} \subset W^{(0)} \subset W^{(1)} \subset \ldots \subset W^{(n)} \subset W^{(n+1)} = W$ where $W^{(\lambda)} = (W^{(\lambda)}; M, M^{(\lambda)})$ is a PL cobordism, $W^{(-1)} \approx M \times I$, $W^{(n+1)}$ -int $W^{(n)} \approx M' \times I$, and $W^{(\lambda)}$ is obtained from $W^{(\lambda-1)}$ by attaching a finite number of λ-handles to $W^{(\lambda-1)}$ in $M^{(\lambda-1)}$, $\lambda = 0, 1, \ldots, n$. That is, each λ-handle $B_i^\lambda \times B_i^{n-\lambda}$ is attached to $M^{(\lambda-1)}$ by a pwl homeomorphism $f_i\colon \dot{B}_i^\lambda \times B_i^{n-\lambda} \longrightarrow M^{(\lambda-1)}$ where the images of the f_i's, for fixed λ, are disjoint in $M^{(\lambda-1)}$. Hence $W^{(\lambda)} = W^{(\lambda-1)} \cup_i f_i(B_i^\lambda \times B_i^{n-\lambda})$ and

$M^{(\lambda)}$ is obtained from $M^{(\lambda-1)}$ by 'surgering out' $\dot{B}_i^\lambda \times B_i^{n-\lambda}$ and replacing with $B_i^\lambda \times \dot{B}_i^{n-\lambda}$. That is, $M^{(\lambda)} = M^{(\lambda-1)} - (\cup_i f_i(\dot{B}_i^\lambda \times B_i^{n-\lambda})) + \cup_i f_i' (B_i^\lambda \times \dot{B}^{n-\lambda})$, where $f_i = f_i' | \dot{B}_i^\lambda \times \dot{B}_i^{n-\lambda}$.

We obtain a dual handle body presentation $W'^{(n+1)} \supset W'^{(n)} \supset \ldots W'^{(0)} \supset W'^{(-1)}$ for $(W; M, M')$ by letting $W'^{(n-\lambda)} = W - \text{int } W^{(\lambda-1)}$, $\lambda = 0, 1, \ldots, n+1$, and $W'^{(n+1)} = W$. For example, $W'^{(-1)} = W - \text{int } W^{(n)} \approx M' \times I$, $W'^{(n)} = W - \text{int } W^{(-1)}$ and $W'^{(n+1)} - \text{int } W'^{(n)} = W^{(-1)} \approx M \times I$: We have the following schematic illustration.

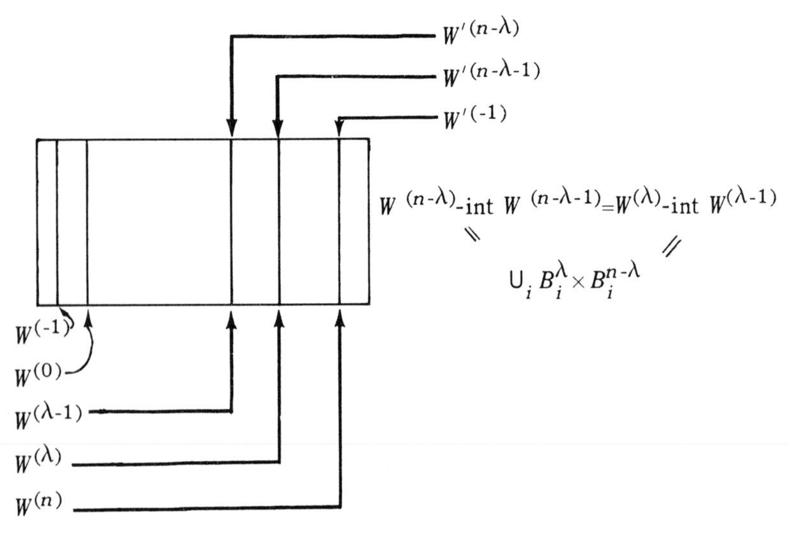

In forming $W'^{(n-\lambda)}$, the $B_i^\lambda \times B_i^{n-\lambda}$ are attached to $\dot{W}'^{(n-\lambda-1)}$ by $B_i^\lambda \times \dot{B}_i^{n-\lambda}$ and in forming $W^{(\lambda)}$, the $B_i^\lambda \times B^{n-\lambda}$ are attached to $\dot{W}^{(\lambda-1)}$ by $\dot{B}_i^\lambda \times B_i^{n-\lambda}$.

If $(W; M, M')$ is a PL cobordism, we can obtain a handlebody presentation as follows. Let T denote the combinatorial triangulation of W and T^2 the second derived of T. Let $\tilde{W}^{(-1)}$ be the regular neighborhood $N(|M''|, T^2)$ of M in T^2 and $\tilde{W}'^{(-1)} = N(|(M')''|, T^2)$. It follows then that $\tilde{W}^{(-1)} \approx M \times I$ and $W'^{(-1)} \approx M' \times I$

(refer to Remark 1, p. 56). Let $\hat{T}$ be the complex $T^2 - \text{int } \tilde{W}^{(-1)} - \text{int } \tilde{W}'^{(-1)}$ where interior is taken in W (e.g., $\text{int } \tilde{W}^{(-1)} = \tilde{W}^{(-1)} - (M \times \{1\})$). Let $\hat{T}^{(i)}$ denote the i-skeleton of $\hat{T}$ and T^4 denote the second barycentric subdivision of T^2. Then $W^{(-1)} = [\tilde{W}^{(-1)}]'' \cup N(|(M \times \{1\})''|, T^4)$, $W^{(0)} = W^{(-1)} \cup N(|(\hat{T}^{(0)})''|, T^4)$ and inductively $W^{(\lambda)} = W^{(\lambda-1)} \cup N(|(\hat{T}^{(\lambda)})''|, T^4)$, $\lambda \leqslant n$. We take $W'^{(-1)}$ to be the closure of $(\tilde{W}'^{(-1)})'' - N(|\hat{T}''|, T^4)$ and $W^{(n+1)} = W^{(n)} \cup W'^{(-1)} = (W'')''$.

Suppose this represents W (under T^2)

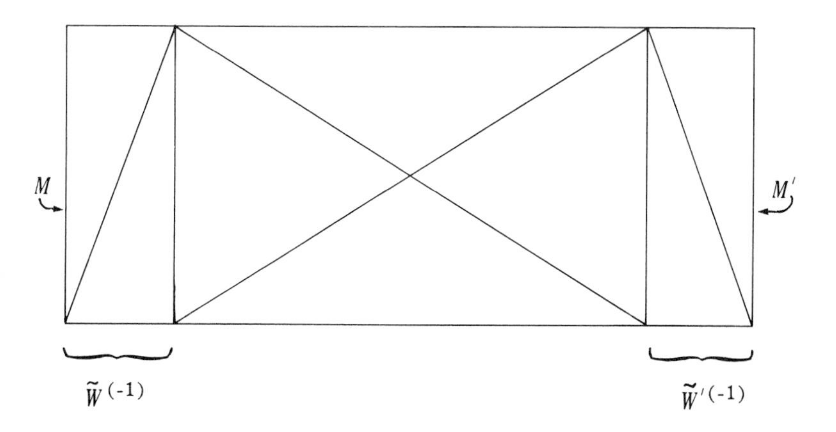

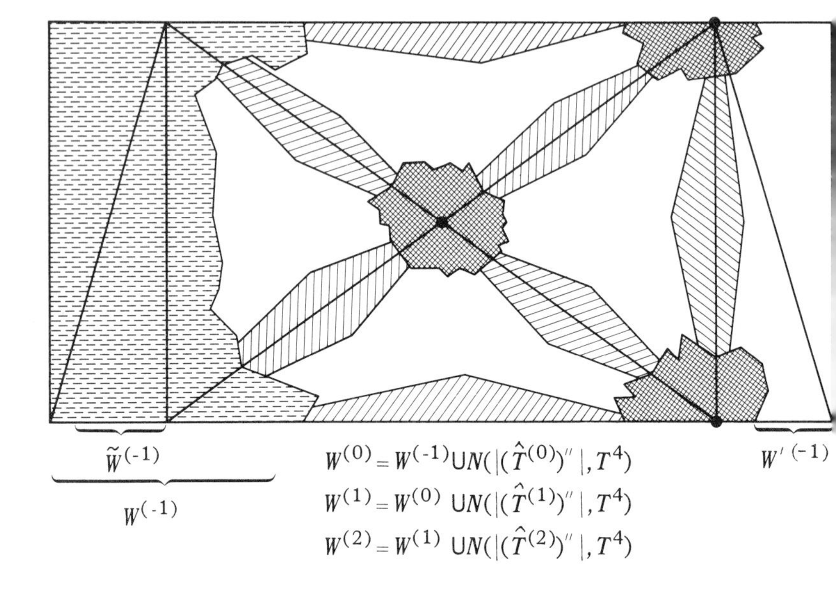

Suppose $(W;M,M')$ is a PL h-cobordism with a handlebody presentation h: $W^{(-1)} \subset W^{(0)} \subset \ldots \subset W^{(n)} \subset W^{(n+1)} = W$. Then the Whitehead torsion $\tau(W,M) = \tau(\mathcal{C}(\hat{W},\hat{M}))$ is defined as discussed earlier. In many instances it becomes useful to define the Whitehead torsion of the pair (W,M) in terms of the handlebody presentation. This allows us to obtain a rather concrete geometrical object in which the relation between $\tau(W,M)$ and the geometry of the situation is more apparent and easier to apply. For example a simple relation between $\tau(W,M)$ and $\tau(W,M')$ can then be established. Also examples of PL h-cobordisms can be obtained having one boundary component as a prescribed $(n\text{-}1)$-manifold M and $\tau(W,M)$ can be made to be any element of $\mathrm{Wh}(\pi_1(M))$ as desired (this will be shown in Theorem I.32).

For $\lambda = 0, \ldots, n$ we now want to consider the pairs $(W^{(\lambda)}, W^{(\lambda-1)})$. $(W^{(\lambda)}, W^{(\lambda-1)})$ retracts by a deformation to $W^{(\lambda-1)} \cup \{$finite number of λ-cells attached along the boundary $M^{(\lambda-1)}\}$. Thus

$$H_i(W^{(\lambda)}, W^{(\lambda-1)}) = \begin{cases} 0 \text{ for } i \neq \lambda \\ \text{free with one generator for each} \\ \lambda\text{-handle for } i = \lambda. \end{cases}$$

We again take $\hat{W}$ to be the universal cover of W and let $\hat{M}$, $\hat{W}^{(\lambda)}$ be $p^{-1}(M)$, $p^{-1}(W^{(\lambda)})$ respectively, where $p: \hat{W} \longrightarrow W$ is the covering map. Then we have that

$$H_i(\hat{W}^{(\lambda)}, \hat{W}^{(\lambda-1)}) = \begin{cases} 0 \text{ for } i \neq \lambda, \\ Z\pi\text{-free with one generator} \\ \text{for each } \lambda\text{-handle of } W^{(\lambda)} \text{ for } i = \lambda. \end{cases}$$

We define $\mathcal{C}^h$ by $C_\lambda^h = H_\lambda(\hat{W}^{(\lambda)}, \hat{W}^{(\lambda-1)})$. Then by making use of the chain complexes $\mathcal{C}^{(\lambda)} = \mathcal{C}(\hat{W}^{(\lambda)}, \hat{M})$ we obtain a filtration of $\mathcal{C}(\hat{W},\hat{M})$: $\mathcal{C}^{(-1)} \subset \mathcal{C}^{(0)} \subset \mathcal{C}^{(1)} \subset \ldots \subset \mathcal{C}(\hat{W},\hat{M})$ (here we will assume that our handlebody presentation is such that $W^{(n)} = W$).

The relative homology groups $H_i(\mathcal{C}^{(\lambda)}/\mathcal{C}^{(\lambda-1)})$ can be identified with $H_i(\hat{W}^{(\lambda)}, \hat{W}^{(\lambda-1)}) = \begin{cases} 0 \text{ for } i \neq \lambda, \\ C_\lambda^h \text{ for } i = \lambda. \end{cases}$

We now can define a new chain complex $\overline{\mathcal{C}}$ by setting $\overline{C}_\lambda = H_\lambda(\mathcal{C}^{(\lambda)}/\mathcal{C}^{(\lambda-1)})$. The boundary homomorphism $\partial: \overline{C}_\lambda \longrightarrow \overline{C}_{\lambda-1}$ is obtained from the exact sequence of the triple $\mathcal{C}^{(\lambda)}, \mathcal{C}^{(\lambda-1)}, \mathcal{C}^{(\lambda-2)}$. The homology groups $H_i(\overline{\mathcal{C}})$ are canonically isomorphic to the groups $H_i(\mathcal{C})$. The proof is similar to that used in showing that the groups $H_i(\mathcal{C}) = H_i(\mathcal{C}(\hat{W},\hat{M}))$ are canonically isomorphic to the groups $H_i(\hat{W},\hat{M})$. Since the proof is short we will indicate it here. The fact that $H_i(\mathcal{C}^{(\lambda)}/\mathcal{C}^{(\mu)}) = 0$ for $i > \lambda$ or $i \leqslant \mu$ is proved by an easy induction on $\lambda - \mu$. Hence $H_i(\mathcal{C}) \cong H_i(\mathcal{C}^{(i+1)}) \cong H_i(\mathcal{C}^{(i+1)}/\mathcal{C}^{(i-2)})$.

Now consider the diagram

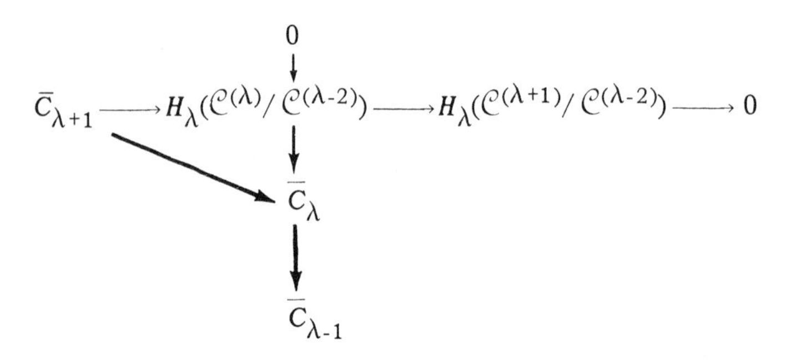

where the vertical line comes from the homology exact sequence of the triple $\mathcal{C}^{(\lambda)}$, $\mathcal{C}^{(\lambda-1)}$, $\mathcal{C}^{(\lambda-2)}$; and the horizontal line from the triple $\mathcal{C}^{(\lambda+1)}$, $\mathcal{C}^{(\lambda)}$, $\mathcal{C}^{(\lambda-2)}$. Inspection shows that the cycle group $\overline{Z}_\lambda$ of $\overline{\mathcal{C}}$ can be identified with $H_\lambda(\mathcal{C}^{(\lambda)}/\mathcal{C}^{(\lambda-2)})$. Hence $H_\lambda(\overline{\mathcal{C}}) = \overline{Z}_\lambda/\overline{B}_\lambda \cong H_\lambda(\mathcal{C}^{(\lambda+1)}/\mathcal{C}^{(\lambda-2)}) \cong H_\lambda(\mathcal{C})$. Thus $H_*(\mathcal{C}) = H_*(\mathcal{C}(\hat{W},\hat{M})) = H_*(\hat{W},\hat{M}) = 0$.

If $\tau(\mathcal{C}^{(\lambda)}/\mathcal{C}^{(\lambda-1)}) = 0$, $\lambda = 0, \ldots, n$, then by an algebraic subdivision theorem that would have been proven if we had developed the algebraic definition of τ completely, we have $\tau(\overline{\mathcal{C}}) = \tau(\mathcal{C}(\hat{W},\hat{M})) = \tau(W,M)$. But since $\tau(\mathcal{C}^{(\lambda)}/\mathcal{C}^{(\lambda-1)}) = \tau(W^{(\lambda)}, W^{(\lambda-1)})$ and each component of $W^{(\lambda)} - W^{(\lambda-1)}$ is simply connected we have that $\tau(W^{(\lambda)}, W^{(\lambda-1)}) = 0$. Now $\overline{\mathcal{C}} = \mathcal{C}^h$ and hence $\tau(\mathcal{C}^h) = \tau(W,M)$. Thus we now may obtain $\tau(W,M)$ in terms of the handlebody presentation by making use of the chain complex $\mathcal{C}^h$ where C_λ^h is $Z\pi$-free with one generator for each λ-handle of $W^{(\lambda)}$.

The following basic theorem has been proved by Mazur, Barden, and Stallings in the smooth category (i.e. with W a differentiable manifold). Therefore, we should apply it only under the additional hypothesis that the PL-manifold W admits a compatible differentiable structure. Whitehead's uniqueness theorem for PL triangulations of differentiable manifolds will then allow us to change the smooth conclusion to a PL one. We will not try to give a proof. (John

Wagoner, in his Ph.D. Thesis at Princeton University, 'Surgery on a map and an approach to the Hauptvermutung', gives in an appendix the necessary machinery and propositions to prove the s-cobordism theorem in the piecewise linear category.)

THEOREM I.30. (the s-cobordism theorem): *If* $(W;M,M')$ *is a* PL *h-cobordism,* dim $W \geqslant 6$, *then* W *is pwl homomorphic to* $M \times I$ *if and only if* $\tau(W,M) = 0$.

COROLLARY I.31. (the h-cobordism theorem): *If* $(W;M,M\ )$ *is a* PL *h-cobordism,* dim $W \geqslant 6$ *and* $\pi_1(M) = \pi_1(M') = 0$ *then* W *is pwl homomorphic to* $M \times I$.

Corollary I.31 was originally proved by Smale in the differentiable case and is just a special case of the s-cobordism theorem since $\tau(W,M) \in \mathrm{Wh}(\pi_1(M)) = 0$. Shortly, as we mentioned earlier, we will actually prove a weak form of the h-cobordism theorem; namely that $W-M' \approx M \times [0,1)$ (here dim $W \geqslant 5$). Also for the weak form, we will not require the additional hypothesis of a compatible differentiable structure and we do not require that $\pi_1(M) = 0$.

THEOREM I.32. *If* M *is a closed connected combinatorial* $(n\text{-}1)$*-manifold,* $n \geqslant 6$, $\pi = \pi_1(M)$ *and* $\tau \in \mathrm{Wh}(\pi)$, *then there exists an h-cobordism* $(W;M,M')$ *with* $\tau(W,M) = \tau$.

Before we give a proof of this theorem we first obtain an easy corollary.

COROLLARY I.33. *If* $(W_1;M_1,M_1')$ *and* $(W_2;M_2,M_2')$ *are* PL *h-cobordisms of* dim $n \geqslant 6$, $M_1 \approx M_2$ *and* $\tau(W_1,M_1), = \tau(W_2,M_2)$, *then* $W_1 \approx W_2$.

Proof. By Theorem I.32 we can construct an h-cobordism $(W;M_1',M')$ with $\tau(W,M_1') = -\tau(W_1,M_1)$. Consider the h-cobordism $(W_1 \cup W;M_1,M')$ formed by identifying M_1' in W_1 with M_1' in W. Then $\tau(W_1 \cup W,M_1) = \tau(W,M_1') + \tau(W_1,M_1) = 0$ and hence by the s-cobordism theorem $W_1 \cup W \approx M_1 \times I$.

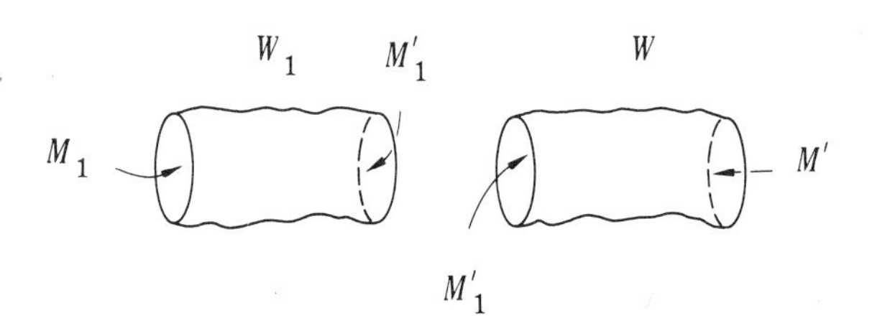

Therefore, there exists a pwl homomorphism φ: $M' \longrightarrow M_1$. Now by hypothesis, there exists a pwl homomorphism $\psi: M_1 \longrightarrow M_2$. Now we consider $W_1 \cup W \cup_{\psi\circ\varphi} W_2$. $W_1 \cup W \cup_{\psi\circ\varphi} W_2 \approx M_1 \times I \cup\ W_2 \approx W_2$. Also since $\tau(W_2,M_2) = \tau(W_1,M_1) = -\ \tau(W,M_1')$ we have that $\tau(W \cup W_2,M_1') = \tau(W_2,M_2) + \tau(W,M_1') = 0$ and as above $W \cup W_2 \approx M_1' \times I$. Thus $W_2 \approx W_1 \cup W \cup W_2 \approx W_1 \cup M_1' \times I \approx W_1$.

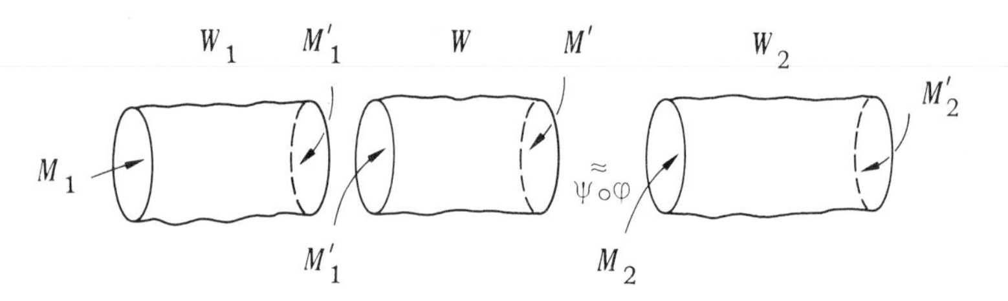

Remark 1. In the above proof we made use of the fact that $M_1 \times I \cup_{\psi} W_2 \approx W_2$ and $W_1 \cup M_1' \times I \approx W_1$. This follows from the observation that if W is a combinatorial manifold with boundary, M is a component of Bd W, and U is a regular neighborhood of M in W, then $U \approx M \times I$. That is, consider the manifold $W \cup (M \times [0,2]) = \widehat{W}$, where $m \in M$ is identified with $(m,0) \in M \times \{0\} \subset M \times [0,2]$. $M \times [0,2]$ can be given a natural triangulation so that $M \times [0,1] \searrow M \times \{0\} = M \subset \mathrm{Bd}\, W$. $\hat{W}$ is clearly a combinatorial manifold with boundary. We just have to check that for any vertex $v \in M$, $\mathrm{st}(v, \hat{W})$ is a combinatorial ball. But $\mathrm{st}(v, \hat{W}) = \mathrm{st}(v, W) \cup \mathrm{st}(v, M \times [0,2])$, where $\mathrm{st}(v, W) \cap \mathrm{st}(v, M \times [0,2]) = \mathrm{st}(v, M)$. Since this is the union of two combinatorial n-balls intersecting in a combinatorial $(n\text{-}1)$-ball, it follows that $\mathrm{st}(v, \hat{W})$ is also a combinatorial n-ball.

Now suppose U is a regular neighborhood of M in W. Then U is a combinatorial n-manifold in $\hat{W}$ so that $U \searrow M$. Hence U is a regular neighborhood of M in $\hat{W}$. Similarly, $M \times [0,1]$ is also a regular neighborhood of $M\ (=M \times [0])$ in $\hat{W}$. Thus $U \approx M \times [0,1]$. That is, $U \approx U'' \approx N(|U''|, \hat{W}'') \approx N(|M''|, \hat{W}'') \approx N(|(M \times I)''|, \hat{W}'') \approx (M \times I)'' \approx M \times I$. The second, third and fourth equivalences follow since each of U, $M \times I$, and M lie in $\mathrm{int}\, \hat{W}$.

Suppose $\psi : M \times [0,1] \to U$ is a pwl homomorphism giving the equivalence $U \approx M \times [0,1]$. From the sequence of equivalences giving $U \approx M \times [0,1]$, it is clear that $\psi(M \times \{1\}) = M$. In many instances it is convenient to note that U is a 'collar' of M in W. That is, there exists a pwl homomorphism φ taking $M \times [0,1]$ onto U so that $\varphi((m,0)) = m$ for all $m \in M$. Let f be the pwl homomorphism of M onto itself defined by the composition

$$M \xrightarrow{(\psi \mid M \times \{1\})^{-1}} M \times \{1\} \xrightarrow{\ p_1\ } M,$$

where $p_1((m,1)) = m$. Define $g : M \times I \to M \times I$ by $g((m,t)) = (f(m), 1\text{-}t)$. Then the desired $\varphi : M \times I \to U$ is given by $\varphi = \psi \circ g$. $\varphi((m,0)) = \psi \circ g((m,0)) = \psi((f(m), 1)) = m$.

Now using the fact that U can be considered as a collar of M in W, it easily follows that $W \cup M \times [0,1] \approx W$. That is, $W = \overline{W - U} \cup U \approx$

$\overline{W-U} \cup \varphi(M\times[0,1]) = \overline{W-U} \cup \varphi(M\times[½,1]) \cup \varphi(M\times[0,½]) \approx$ $\overline{W-U} \cup \varphi(M\times[0,1]) \cup M\times[0,1] \approx W \cup M\times[0,1]$. We note that if U is a small enough regular neighborhood of M in W so that $\overline{W-U}$ is also a combinatorial n-manifold with boundary, then $\overline{W-U} \approx W$. For $\overline{W-U} \approx \overline{W-U} \cup \hat{M}\times I$ (where $\hat{M} = \overline{W-U} \cap M\times I) \approx \overline{W-U} \cup U \approx W$.

Remark 2. The above observations also allow us to obtain a proof of Corollary I.33 without making use of the existence theorem, if we assume the result that $W-M' \approx M\times[0,1))$ (Theorem I.34). That is, consider W_1, W_2 as given in Corollary I.33. Now $W_1 \approx \overline{W_1-U'} \subset W_1-M_1' \approx M_1\times[0,1)$, where U_1' is a collar of M_1' in W_1. Let U_2 be a collar of M_2 in W_2. Since $U_2 \approx M_2\times I$, $M_1 \approx M_2$ and $W_1 \overset{e}{\subset} M_1\times[0,1)$, it follows that W_1 can be pwl embedded in $U_2 \subset W_2$, so that under the embedding $W_1 \overset{e}{\subset} U_2 \subset W_2$, M_1 is carried pwl onto M_2. That is, there are subdivision αW_1 and βW_2 so that we can consider αW_1 as a subcomplex of βW_2 with $\alpha M_1 = \beta M_2$. Now $\tau(\beta W_2, \beta M_2) = \tau(\alpha W_1, \alpha M_1) + \tau(\overline{\beta W_2-\alpha W_1}, \tilde{M})$, where $\tilde{M} = \overline{\beta W_2-\alpha W_1} \cap \alpha W_1$. Since $\tau(\beta W_2, \beta M_2) = \tau(W_2, M_2) = \tau(W_1, M_1) = \tau(\alpha W_1, \alpha M_1)$, it follows that $\tau(\overline{\beta W_2-\alpha W_1}, \tilde{M}) = 0$. But then by the s-cobordism theorem $\beta W_2-\alpha W_1 \approx \tilde{M}\times I$. Thus $W_2 \approx \beta W_2 \approx \alpha W_1 \cup \beta W_2-\alpha W_1 \approx \alpha W_1 \cup \tilde{M}\times I \approx \alpha W_1 \approx W_1$.

We now want to give a proof of Theorem I.32. At one step in the proof it will be necessary to use a result from differential topology, so here also we assume that M admits a compatible differentiable structure. In order to outline the proof, we need to make use of the concept of attaching handles to manifolds as discussed earlier. We first review this concept briefly.

Let V be a PL n-manifold with non-empty boundary Bd V. Consider the n-cube $D^n = [-1,1]^n$ as a product $D^p\times D^q$, $p+q = n$, and suppose g: (Bd $D^p) \times D^q \longrightarrow$ Bd V is a PL imbedding. Then the PL n-manifold $V_1 = V \cup_g (D^p \times D^q)$ is said to be obtained by attaching a p-handle to V, or a handle of index p. We sometimes refer to g as the attaching map and to the (p-1)-sphere g(Bd $D^p \times 0) \subset$ Bd V as the attaching sphere.

We note that a p-handle and a $(p+1)$-handle can be attached in succession to V is such a way as to cancel each other, i.e., so that the resulting manifold is equivalent to V. This idea is at the heart of the proof of the s-cobordism theorem and h-cobordism theorem. That is, suppose (W^n, M, M') with $n > 5$ is a given h-cobordism. It is first proved that W can be obtained from $M \times [0,1]$ by first attaching p-handles $(2 \leqslant p \leqslant n-3)$ to $M \times I$ along its 'right' boundary $M \times \{1\}$ to obtain a manifold W_1, and then attaching $(p+1)$-handles to W_1 along its 'right' boundary. In short, W can be constructed by attaching h handles of two adjacent indexes to $M \times I$. Considering the universal cover $\hat{W}$ of W we have $\mathcal{C}^h$: $H_{p+1}(\hat{W}, \hat{W}_1) \longrightarrow H_p(\hat{W}_1, \hat{M} \times I) \longrightarrow 0$ where $H_{p+1}(\hat{W}, \hat{W}_1)$ is $Z\pi$ free with one generator corresponding to each of the $(p+1)$-handles attached to the 'right' boundary of W_1 and $H_p(\hat{W}_1, \widehat{M \times I})$ is $Z\pi$-free with one generator corresponding to each of the p-handles attached to $M \times \{1\} \subset M \times I$. We have then, in terms of these generators, the matrix of the boundary homomorphism $\partial: H_{p+1}(\hat{W}, \hat{W}_1) \longrightarrow H_p(\hat{W}_1, \widehat{M \times I})$ is non-singular $k \times k$ matrix (a_{ij}) which represents the torsion $\tau(W,M) \in \mathrm{Wh}(\pi_1(M))$. The assumption that $\tau(W,M) = 0$ implies that (a_{ij}) can be reduced to the indentity matrix by appropriate row operations, and this in turn implies that the p-handles and $(p+1)$-handles can be rearranged so as to cancel each other in pairs, proving that $W \approx M \times I$.

Exercise

Let M be a closed PL $(n\text{-}1)$-manifold and let $W_1 = (M \times I) \cup_g (D^p \times D^q)$ be constructed by attaching a p-handle to the right boundary $M \times \{1\}$ of $M \times I$. If M_1 denotes the right boundary of W_1, then show W_1 can also be constructed by attaching a q-handle to the left boundary $M_1 \times \{0\}$ of $M_1 \times I$.

Hint: If $D_0^q = [-½, ½]^q \subset \text{int } D^q$, then $W_1 - [(D^p \times D_0^q) \cup_g (\text{Bd } D^p \times D_0^q) \times I]$ is PL homomorphic to $M_1 \times I$.

Proof of the existence theorem (Theorem I. 32): Let (a_{ij}) be a non-singular $k \times k$ matrix over $Z\pi$ which represents the given $\tau \in \text{Wh}(\pi_1(M))$. In $M \times \{1\}$, the right boundary of $M \times I$ choose k disjoint $(n\text{-}1)$-cells $B_1, \ldots, B_k$. Attach k 2-handles $D_i^2 \times D_i^{n-2}$ by homomorphisms g_i: $(\text{Bd } D_i^2 \times D_i^{n-2}) \longrightarrow \text{int } B_i$ so that the resulting manifold $W_1 = (M \times I) \cup (\cup_{i=1}^k g_i\ D_i^2 \times D_i^{n-2})$ can also be described as that obtained from $M \times I$ by pasting k copies of $S^2 \times D^{n-2}$ along $(n\text{-}1)$-cells in its right boundary. We note $\pi_1(W_1) = \pi_1(M \times I) = \pi_1(M)$. Denote by M_1 the right hand boundary of the resulting PL n-manifold W_1.

Assertion: Each element $\pi_2(W_1, M)$ can be represented by a 2-sphere $S \subset M_1$, such that S has a PL product neighborhood $S \times D^{n-3}$ in M_1.

Proof of the assertion: We first show that the composition $\pi_2(M_1) \longrightarrow \pi_2(W_1) \longrightarrow \pi_2(W_1, M)$ of homomorphisms is onto $\pi_2(W_1, M)$. That is, since $\pi_1(M) = \pi_1(W_1)$, $\pi_2(W_1) \longrightarrow \pi_2(W_1, M)$ is onto by exactness. Now by the above exercise, we can regard W_1 as having been obtained from $M_1 \times I$ by attaching k handles of index $n\text{-}2 > 2$ along its left boundary. Hence it follows that

$\pi_2(M_1) \longrightarrow \pi_2(W_1)$ is onto also. Thus each element $\alpha \in \pi_2(W_1,M)$ can be represented by a map $f\colon S^2 \longrightarrow M_1$. Since dim $M_1 = n\text{-}1 \geqslant 5$, we may assume by general position that f is a PL imbedding. To show that $S \subset M_1$ can also be chosen so as to have a product neighborhood is where the differentiability hypothesis is needed. That is, by machinery of differential topology, we can choose S so that its normal bundle is trivial.

Now is $(\hat{W}_1, \hat{M})$ is the universal cover of (W_1,M), then $\pi_2(W_1,M) \cong \pi_2(\hat{W}_1,\hat{M})$ and $\pi_1(\hat{W}_1) = \pi_1(\hat{M}) = 0$; hence $\pi_2(\hat{W}_1,\hat{M}) = H_2(\hat{W}_1,\hat{M})$ by the relative Hurewicz Theorem. Since each of the 2-handles attached to $M \times I$ lifts to a collection of 2-handles, attached to $\widehat{M \times I}$, which are transitively permuted by the covering transformations, and since $\pi_2(W_1,M) \cong H_2(\hat{W}_1,\hat{M})$, it is clear that $\pi_2(W_1,M)$ is free over $Z\pi_1(M)$ with generators $\alpha_1, \ldots, \alpha_k$, one corresponding to each 2-handle. By the above assertion, the elements $\beta_i = \sum_{i=1}^{k} a_{ij}\alpha_j \in \pi_2(W_1,M)$ (i $1, \ldots, k$) can be represented by mutually disjoint 2-spheres $S_1, \ldots, S_k$ in M_1. We now use $S_1, \ldots, S_k$ as attaching spheres to attach k 3-handles to W_1 along M_1, obtaining the required manifold W. Let M' be its right boundary.

We must now verify that $(W;M,M')$ is indeed an h-cobordism with torsion τ. Clearly $\pi_1(M) \cong \pi_1(W_1) \cong \pi_1(W)$, and $\pi_1(M') \cong \pi_1(W)$ since W can be obtained from $M' \times I$ by attaching handles of indexes $n\text{-}3 \geqslant 3$ and $n\text{-}2 \geqslant 4$.

We now need to consider some of the concepts discussed earlier. That is, W has a handlebody decomposition $h\colon W^{(-1)} = W^{(0)} = W^{(1)} = M \times I$, $W^{(2)} = W_1$, $W^{(3)} = \ldots W^{(n)} = W$ and we consider the chain complex $\mathcal{C}^h$ where $C_i^h = H_i(W^{(i)}, W^{(i-1)})$. We then have that $H_*(\hat{W},\hat{M}) \cong H_*(\mathcal{C}(\hat{W},\hat{M})) \cong H_*(\mathcal{C}^h)$ and $\tau(\mathcal{C}^h) = \tau(W,M)$. Clearly $C_3^h \cong 0$ if $i \neq 2,3$ and $C_2^h \cong H_2(\hat{W}_1, \hat{M} \times I)$ and $C_3^h \cong H_3(\hat{W},\hat{W}_1)$ are both $Z\pi$-free modules with k generators. Thus the chain complex $\mathcal{C}^h$

contains only two adjacent non-trivial modules and has the form: $0 \to C_3^h \xrightarrow{\delta} C_2^h \to 0$. C_2^h has generators $c_2 = (\alpha_1, \ldots, \alpha_k)$, where each α_i corresponds to a 2-handle attached to $M \times I$ in forming W_1. C_3^h has generators $c_3 = (\beta_1, \ldots, \beta_k)$, where each β_i corresponds to a 3-handle attached to W_1 in forming W. Also because of the manner in which we attached the various 3-handles to W_1, the matrix of ∂ is the matrix (a_{ij}) with which we started. That is, $\partial \beta_i = \Sigma_{j=1}^k a_{ij} \alpha_j$, $i = 1, \ldots, k$. Now $\tau(\mathcal{C}^h) = \Sigma(-1)^i [b_i h_i b_{i-1}/c_i]$. We choose the basis $b_2 = (\alpha_1, \ldots, \alpha_k)$ for $B_2^h = C_2^h$. Then $\bar{b}_2 = (\partial^{-1}\alpha_1, \ldots, \partial^{-1}\alpha_k)$ gives a new basis for C^h and $\partial^{-1}\alpha_i = \Sigma_{j=1}^k \bar{a}_{ij}\beta_i$, where $(\bar{a}_{ij}) = (a_{ij})^{-1}$. Thus $\tau(\mathcal{C}^h) = (-1)^3 [\bar{b}_2/c_3] + (-1)^2 [c_2/c_2] = (-1)^3 [(\bar{a}_{ij})] + (-1)^2 [I] = (-1)^2 [(a_{ij})]$ and $\tau(\mathcal{C}^h) = \tau(W,M)$ is represented by the matrix (a_{ij}) as desired.

Finally since (a_{ij}) is non-singular and ∂ is an isomorphism we have that $H_*(\mathcal{C}^h) = 0$. Thus $0 = H_*(\mathcal{C}^h) \cong H_*(\hat{W}, \hat{M}) \cong \pi_*(\hat{W}, \hat{M}) \cong \pi_*(W,M)$ and M is a deformation retract of W. By duality $H_*(\hat{W}, \hat{M}') = 0$ so that M is also a deformation retract of W. Therefore $(W;M,M')$ is an h-cobordism.

THEOREM I.34. *If $(W;M,M')$ is a PL h-cobordism where $\dim W = n \geqslant 5$, then $W - M' \approx M \times [0,1)$ (similarly, $W - M \approx M' \times [0,1)$).*

LEMMA FOR THEOREM I.34. *Suppose $(W;M,M')$ is a PL h-cobordism where $\dim W = n \geqslant 5$, U is a neighborhood of M in W and U' is a neighborhood of M' in W with $U \cap U' = \emptyset$ such that there is a pwl homomorphism φ taking $M \times [0,1)$ onto U with $\varphi(m,0) = m$ for $m \in M$ and there is a pwl homomorphism φ' taking $M' \times [0,1)$ onto U' with $\varphi'(m',0) = m'$ for $m' \in M'$. Then given $0 < \delta < 1$, $0 < \delta' < 1$ there exists a pwl homomorphism $h: W \longrightarrow W$ such that $h(\varphi(M \times [0,1))) \cup \varphi'(M' \times [0,1)) = W$ and $h =$ identity on $\varphi(M \times [0,1-\delta)) \cup \varphi'(M' \times [0,1-\delta'))$.*

Proof. Since $(W;M,M')$ is an h-cobordism and $N = W - \varphi(M \times [0,1-\delta)) - \varphi'(M' \times [0,1-\delta')) \approx W$, it follows that $\pi_i(W,M) = \pi_i(W, \varphi(M \times [0,1))) = \pi_i(N, \varphi(M \times [1-\delta,1))) = 0$ for all i. Similarly $\pi_i(N, \varphi'(M' \times [1-\delta',1))) = 0$ for all i. Also then $\pi_i(\mathring{N}, \varphi(M \times (1-\delta,1))) = \pi_i(\mathring{N}, \varphi'(M' \times (1-\delta ,1))) = 0$ for all i. Let T be a locally finite triangulation of $\mathring{N}$ so that $\mathring{N} - \varphi(M \times (1-\delta,1-\delta/2)) - \varphi'(M' \times (1-\delta',1-\delta'/2)) = N_1$ is a subcomplex under T. Let $\hat{K}$ be the $(n-3)$-skeleton of N_1 and $K = \varphi(M \times (1-\delta,1-\delta/2]) \cup \hat{K}$. Let L be the subcomplex of the first barycentric subdivision T' of T maximal with respect to missing K. Then every n-simplex $\sigma \in T'$ either lies in K' or in L or is the join of a simplex in K' of L. Also there are only a finite number of n-simplexes of the third type. Now $\varphi(M-(1-\delta,1))-K$ and $\varphi'(M' \times (1-\delta ,1))-L$ are each compact and hence we can apply the engulfing lemma. Therefore, there exists $h_1 : \mathring{N} \longrightarrow \mathring{N}$ such that $h_1(\varphi(M \times (1-\delta,1))) \supset K$ and $h_2 : \mathring{N} \longrightarrow \mathring{N}$ such that $h_2(\varphi'(M' \times (1-\delta',1))) \supset L$. Also since each of these are the identity outside a compact set, they both extend to all of W by the identity on $\varphi(M \times [0,1-\delta]) \cup \varphi'(M' \times [0,1-\delta'])$. We denote the extensions by h_1 and h_2 also. Now because of the definitions of K and L we can apply Lemma IV.12 and hence there exists $h_3 : \mathring{N} \longrightarrow \mathring{N}$ such that $h_1(\varphi(M \times (1-\delta,1))) \cup h_3 h_2(\varphi'(M' \times (1-\delta',1))) = N$. Since h_3 = identity on $K \cup L$ we can also extend h_3 by the identity to all of W. Thus we have three pwl homomorphism h_1, h_2, and h_3 taking $W \longrightarrow W$ each the identity on $\varphi(M \times [0,1-\delta]) \cup \varphi' (M' \times [0,1-\delta'])$. The h promised above is then $h_2^{-1} \circ h_3^{-1} \circ h_1$.

Proof of Theorem I.34. The U and U' of the previous lemma are obtained by considering the interior of sufficiently small regular neighborhoods of M and M' in W respectively. For convenience we will call these neighborhoods $\varphi(M \times [0,1))$ and $\varphi'(M' \times [0,1))$

respectively. Applying the lemma we obtain a pwl homeomorphism $h_1: W \longrightarrow W$ such that $h_1 \varphi(M \times [0,1)) \cup \varphi'(M' \times [0,1)) = W$ and $h_1 =$ identity on $\varphi(M \times [0,1-\frac{1}{2})) \cup \varphi'(M' \times [0,1-\frac{1}{2}))$. Now there exists a $\delta_1, 0 < \delta_1 < \frac{1}{2}$, so that $h_1(\varphi(M \times [0,1))) \cap \varphi'(M' \times [0,\delta_1)) = \emptyset$. Now we apply the previous lemma again where U is now $(\varphi(M \times [0,1)))$ and $U' = \varphi'(M' \times [0,\delta_1))$. That is, there exists a pwl homomorphism $h_2: W \longrightarrow W$ such that $h_2 h_1(\varphi(M \times [0,1))) \cup \varphi'(M' \times [0,\delta_1)) = W$ and $h_2 =$ identity on $h_1(\varphi(M \times [0,1-\frac{1}{4}))) \cup \varphi(M \times [0,\delta_1/2))$. In a similar fashion we obtain an h_3 so that $h_3(h_2 h_1 \varphi(M \times [0,1))) \cup \varphi'(M' \times [0,\delta_2)) = W$, $\delta_2 < \frac{1}{4}$, and so that $h_3 =$ identity on $h_2 h_1 \circ \varphi(M \times [0,1-1/8))$ $\cup \varphi'(M' \times [0,\delta_2/2))$. By repeating this process we inductively define pwl homomorphisms $h_j: W \longrightarrow W$ such that $h_j(h_{j-1} \circ \ldots \circ \varphi(M \times [0,1)))$ $\cup \varphi'(M' \times [0,\delta_{j-1})) = W$, $\delta_{j-1} < \frac{1}{2}^{j-1}$ and $h_j =$ identity on $h_{j-1} \circ \ldots \circ h_2 \circ h_1 \circ \varphi(M \times [0,1-1/2^j)) \cup \varphi'(M' \times [0,\delta_{j-1}/2))$. Thus each h_j expands the previous image of $\varphi(M \times [0,1))$ towards M' keeping the previous image fixed on increasingly larger subsets of it. Let $k_j = h_j \circ h_{j-1} \circ \ldots \circ h_2 \circ h_1 \circ \varphi$ and $k = \lim j \rightarrow \infty\ k_j$. Then k is a pwl homeomorphism taking $M \times [0,1)$ onto $W - M'$. k is a homomorphism and is pwl since k_j has these properties and each point in $M \times [0,1)$ is only moved by a finite number of k_j's. k is onto since $k_j(M \times [0,1)) \cup \varphi'(M' \times [0,\delta_{j-1})) = W$ and $\delta_{j-1} < \frac{1}{2}^{j-1}$.

COROLLARY I.35. *If* $(W;M,M')$ *is a* PL *h-cobordism where* dim $W = n \geqslant 5$ *and* xM' *is the cone over* M' *and* yM *is the cone over* M, *then* $W \cup xM$ *is topologically homomorphic to* yM.

Proof. $yM - \{y\} \approx M \times [0,1) \approx W - M' \approx W \cup M' \times [0,1) \approx W \cup xM' - \{x\}$. Thus $yM = M \times [0,1]/M \times \{1\}$ is homomorphic to $W \cup M' \times [0,1]/M' \times \{1\} = W \cup xM'$.

COROLLARY I.36. *The Poincaré conjecture via the weak form of the h-cobordism* (Theorem I.34).

Proof. Suppose N is a combinatorial closed homotopy n-sphere, $n \geqslant 5$. Let σ_1, σ_2 be two disjoint n-simplexes in N. Then $N - \sigma_1 \approx N - \{\text{point}\}$ and hence if we show that $N - \sigma_1 \approx E^n$, it follows that $N \underset{T}{=} S^n$. Now $N - \text{int}\, \sigma_1 - \text{int}\, \sigma_2 = W$ is an h-cobordism $(W; \dot{\sigma}_2, \dot{\sigma}_1)$. Thus $\dot{\sigma}_2 \times [0,1) \approx W - \dot{\sigma}_1 = N - \text{int}\, \sigma_2 - \sigma_1$ and hence $E^n \approx \sigma_2 \cup (N - \text{int}\, \sigma_2 - \sigma_1) = N - \sigma_1$.

LEMMA I.37. *If K is a complex, xK is the cone over K, and α is any subdivision of xK so that $|\alpha K| \cap |\text{st}(x, \alpha xK)| = \emptyset$, then $\alpha(xK) - \overset{\circ}{\text{st}}(x, \alpha xK) \approx K \times I$.*

Proof. The proof will be by induction on the dimension of K. It is clearly trivial if $\dim K = 0$.

Let us assume the result for all complexes $\hat{K}$, where $\dim \hat{K} \leqslant k-1$ and suppose $\dim K = k$. First, we note that the result is true for a k-simplex Δ. That is, suppose α is a subdivision of $x\Delta$ with $|\alpha\Delta| \cap |\text{st}(x, \alpha x \Delta)| = \emptyset$. By the inductive hypothesis $\alpha x \dot{\Delta} - \overset{\circ}{\text{st}}(x, \alpha x \dot{\Delta}) \approx \dot{\Delta} \times I$. Now $\alpha(x\Delta)$ is just a subdivision of a $(k+1)$-simplex, hence $\text{lk}(x, \alpha x\Delta)$ is a combinatorial k-ball. Thus $\alpha x\Delta - \overset{\circ}{\text{st}}(x, \alpha x\Delta) \approx (\alpha x\Delta - \overset{\circ}{\text{st}}(x, \alpha x\Delta)) \cup \text{st}(x, \alpha x\Delta) = \alpha x\Delta$, and hence $\alpha x\Delta - \overset{\circ}{\text{st}}(x, \alpha x\Delta)$ is a combinatorial $(k+1)$-ball. Now $\alpha\Delta \cup \alpha x\dot{\Delta} - \overset{\circ}{\text{st}}(x, \alpha x\dot{\Delta}) \cup \text{lk}(x, \alpha\Delta) \approx \Delta \times \{0\} \cup \dot{\Delta} \times I \cup \Delta \times \{1\} = (\Delta \times I)$, and hence we have a pwl homeomorphism defined on the boundary of the combinatorial $(k+1)$-ball, $\alpha x\Delta - \overset{\circ}{\text{st}}(x, \alpha x\Delta) = B^{k+1}$, to the boundary of the combinatorial $(k+1)$-ball $\Delta \times I$. Hence $B^{k+1} \approx b\dot{B}^{k+1} \approx b(\Delta \times I)^{\cdot} \approx \Delta \times I$, where the sequence of combinatorial equivalences preserves the given equivalence $\dot{B}^{k+1} \approx (\Delta \times I)^{\cdot}$.

Now we consider αxK and let $K^{(k-1)}$ be the $(k$-$1)$-skeleton of K. αxK induces a subdivision $\alpha xK^{(k-1)}$ and hence the inductive hypothesis applies to give $\alpha(xK^{(k-1)}) - \overset{\circ}{\text{st}}(x, \alpha xK^{(k-1)}) \approx K^{(k-1)} \times I$. Consider any k-simplex $\Delta \in K$, then the above equivalence defines an equivalence $\alpha x\dot{\Delta} - \overset{\circ}{\text{st}}(x,\alpha x\dot{\Delta}) \approx \dot{\Delta}\times I$. We now extend this as done above to $\alpha x\Delta - \overset{\circ}{\text{st}}(x,\alpha x\Delta) \approx \Delta\times I$. Repeating this for each k-simplex of K gives the desired result.

PROPOSITION I.38. *If $(W;M,M')$ is a PL h-cobordism and if $W \cup xM'$ is pwl homomorphic to yM, then W is pwl homomorphic to $M\times I$.*

Proof. Let $\varphi\colon yM \longrightarrow W\cup xM'$ be a pwl homomorphism and β, α subdivisions so that $\varphi\colon \beta(yM) \longrightarrow \alpha(W\cup xM')$ is an isomorphism. Also suppose that β and α are fine enough subdivisions so that $|\text{st}(y, \beta(yM))| \cap |M| = \emptyset$ and $|\text{st}(x,\alpha(W\cup xM'))| \cup |W| = \emptyset$. By the previous lemma $\beta(yM) - \overset{\circ}{\text{st}}(y, \beta(yM)) \approx M\times I$. Also we may suppose that $\varphi(y) = x$. For if yM and $W\cup xM'$ are both manifolds, then since both x and y are interior points, we may assume $\varphi\colon yM \longrightarrow W\cup xM'$ is a pwl homomorphism carrying y to x. Otherwise, since $yM - \{y\} \approx M\times[0,1)$ and $(W\cup xM') - \{x\} \approx W\cup M'\times[0,1)$ are both manifolds, if yM and $W\cup xM'$ are both not manifolds, then we must have $\varphi(y) = x$. Thus $M\times I \approx \beta(yM) - \overset{\circ}{\text{st}}(y,\beta(yM)) \cong \alpha(W\cup xM') - \overset{\circ}{\text{st}}(x,\alpha(W\cup xM')) \approx W\cup M'\times I \approx W$.

THEOREM I.39. (Counterexample to the Hauptvermutung) *For $n\geqslant 6$, there exists complexes K and L such that K is homomorphic to L, but $K \not\approx L$.*

Proof. For $n\geqslant 6$, let M be a closed connected combinatorial $(n$-$1)$-manifold such that $\pi = \pi_1(M) \neq 0$ and so that $\text{Wh}(\pi) \neq 0$. Let $\tau \neq 0 \in \text{Wh}(\pi)$. Then there exists a Pl h-cobordism $(W;M,M')$ with $\tau(W,M) = \tau$. Let $K = W\cup xM'$ and $L = yM$. Then K is topologically homomorphic with L by Corollary I.35.

Now by the s-cobordism theorem, since $\tau(W,M) = \tau\neq 0$, $W \not\approx M\times I$ and by the above proposition the, $K = W\cup xM' \not\approx yM = L$. (Actually we are only using the easy part of the s-cobordism theorem, namely if $W \approx M \times I$ then $\tau(W,M) = 0$. For if $W \approx M \times I$, then some $\alpha W \cong \beta(M \times I)$ and some $\delta(M \times I) \cong \gamma\beta(M \times I)$. Now since $M \times I \searrow M$ we have $\delta(M \times I) \searrow \delta M$ and hence $\gamma\alpha W \searrow \gamma\alpha M$. Thus $\tau(\gamma\alpha W, \gamma\alpha M) = 0$ and since Whitehead torsion is invariant under subdivision $\tau(W,M) = 0$. Therefore if $\tau(W,M) \neq 0$, then $M \times I \not\approx W$. Also we noted earlier that if T_5 is the cyclic group of order 5 then $\mathrm{Wh}(T_5) \cong \mathbf{Z} \neq 0$. Hence taking K to be the 2-complex formed by attaching a disk D to a loop l by the formula l^5, then $\pi_1(K) \cong T_5$ and M is just the boundary of a regular neighborhood of K in E^n $(n \geqslant 6)$. We recall that if t is the generator of T_5, taking $t + t^{-1} + 1 \in ZT_5$, the class of $(t + t^{-1} + 1)$ in $\mathrm{Wh}(T_5)$ is the generator of $\mathrm{Wh}(T_5)$.)

We will now give an alternative method of obtaining Theorem I.34 for $n \geqslant 6$, the proof of which is of interest in itself. We first give some definitions and notation. Let $C = (W;M,M')$ and $\overline{C} = (\overline{W};\overline{M},\overline{M}')$ be two PL h-cobordisms and $f\colon M' \longrightarrow \overline{M}$ a pwl homomorphism. We define the h-cobordism $C \cup_f \overline{C} = (W \cup \overline{W}; M, \overline{M}')$ by 'gluing' the right end M' of C to the left end $\overline{M}$ of $\overline{C}$ by means of f.

The h-cobordism $C = (W;M,M')$ is said to invertible if and only if there exists an h-cobordism $\overline{C} = (\overline{W};\overline{M},\overline{M}')$ and PL homomorphisms $f\colon M' \longrightarrow \overline{M}$ and $g\colon \overline{M}' \longrightarrow M$ such that the compositions $C \cup_f \overline{C}$ and $\overline{C} \cup_g C$ are both trivial (i.e. $C \cup_f \overline{C} \approx M \times I$ and $\overline{C} \cup_g C \approx \overline{M} \times I$).

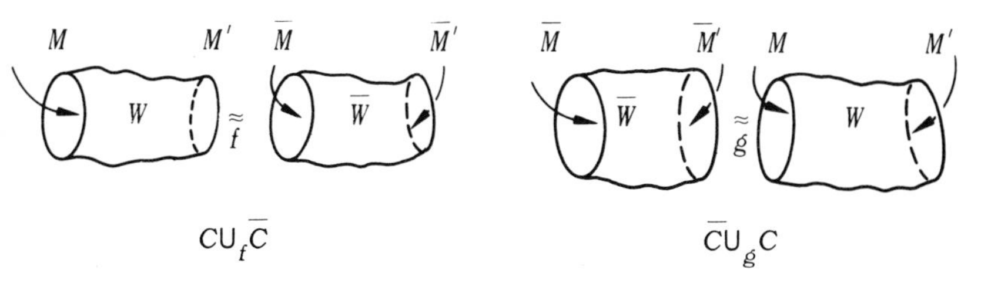

PROPOSITION I (Stallings). *If the h-cobordism* $(W;M,M')$ *is invertible, then* $W - M'$ *is* PL *homomorphic to* $M \times [0,1) \approx M \times [0,\infty)$.

COROLLARY I.35 (I) *If* $(W;M,M')$ *is an invertible h-cobordism, then* $W \cup xM'$ *is topologically homomorphic to* yM.

THEOREM I.34 (I) *If* $(W;M,M')$ *is a* PL *h-cobordism where* dim $W = n \geqslant 6$, *then* $W - M \approx M \times [0,1)$.

The proof of Corollary I.35 (I) is the same as that of Corollary I.35 because of Proposition I. Theorem I.34 (I) follows from Proposition I, the existence theorem (Theorem I.32), and the s-cobordism theorem (Theorem I.30). For if $C = (W;M,M')$ is an h-cobordism with $\tau(C) = \tau_0 \in \mathrm{Wh}(\pi_1(M'))$, the existence theorem gives us an h-cobordism $\overline{C} = (\overline{W};M',\overline{M}')$ with $\tau(\overline{C}) = -\tau_0$. Then $\tau(C \cup \overline{C}) = \tau_0 - \tau_0 = 0$, and hence $C \cup \overline{C}$ is trivial by the s-cobordism theorem. Hence $\overline{M}' \approx M$, and similarly $\overline{C} \cup C$ is trivial. Thus $(W;M,M')$ is invertible and Proposition I applies.

Proof of Proposition I: Since $(W;M,M')$ is invertible, there exists an 'inverse' h-cobordism $(\overline{W}:\overline{M},\overline{M}')$ and $f\colon M' \longrightarrow \overline{M}$ and $g\colon \overline{M}' \longrightarrow M$ such that $W \cup_f \overline{W} \underset{h}{\approx} M \times I$ and $\overline{W} \cup_g W \underset{k}{\approx} \overline{M} \times I \approx M' \times I$. Consider the product $M \times [0,\infty)$. We want to define for each i $0,1,2,\ldots$, a PL homomorphism $h_i\colon W \cup_f \overline{W} \longrightarrow M \times [i, i+1]$ so that in writing $W = h_i(W)$ and $\overline{W}_i = h_i(\overline{W})$ we will have $\overline{W}_{i-1} \cup W_i \approx M' \times I$. We will then have the following schematic picture of $M \times [0,\infty)$:

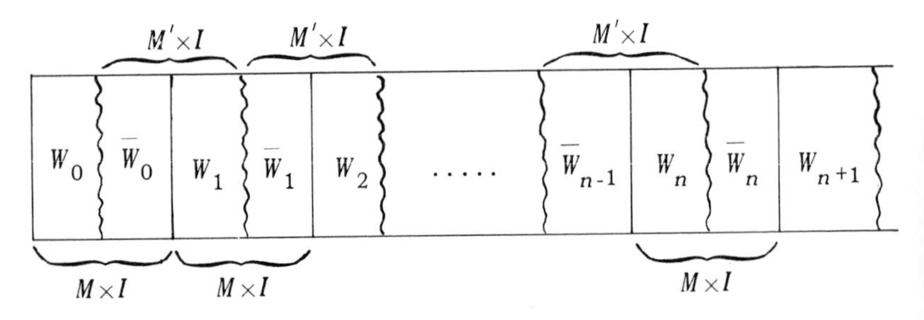

First pick h_0: $W \cup_f \overline{W} \longrightarrow M \times [0,1]$ arbitrarily (e.g. $h_0 = h$). Then suppose inductively that $h_0, \ldots, h_{n-1}$ have been defined in such a way that each $\overline{W}_{i-1} \cup W_i$ $(i \leqslant n-1)$ is PL homomorphic to $M' \times I$. Now since any PL homomorphism of M onto $M \times \{n\}$ can be extended to a PL homomorphism of $W \cup_f \overline{W}$ onto $M \times [n, n+1]$, we can use g to define the PL homomorphism h_n: $W \cup_f \overline{W} \longrightarrow M \times [n, n+1]$ so that $\overline{W}_{n-1} \cup W_n$ is PL homomorphic to $M' \times I$. That is, we have $h_{n-1}|_{\overline{M}'}$: $\overline{M}' \longrightarrow M \times \{n\}$ and using this with g^{-1} we get $h_{n-1}|_{\overline{M}'} \circ g^{-1}$: $M \longrightarrow M \times \{n\}$. Therefore $h_{n-1}|_{\overline{M}'} \circ g^{-1}$ can be extended to a PL homomorphism of $W \cup_f \overline{W} \longrightarrow M \times [n, n+1]$, which is our desired h_n. Now $M' \times I \approx \overline{M} \times I \underset{k^{-1}}{\approx} W \cup_g W$ and we can define a PL homomorphism $\overline{h}_n$ of $\overline{W} \cup_g W$ onto $\overline{W}_{n-1} \cup W_n$ by $\overline{h}_n(x) = h_{n-1}(x)$ for $x \in \overline{W}$ and $\overline{h}_n(x) = h_n(x)$ for $x \in W$. We note that for $x \in \overline{M}'$, x is identified with $g(x) \in \overline{M}'$. But $\overline{h}_n(x) = h_{n-1}(x)$ and $\overline{h}_n(g(x)) = h_{n-1}(x)$, so x and $g(x)$ are carried by $\overline{h}_n$ to the same point of $\overline{W}_{n-1} \cup W_n$. This completes the induction,

Now it is clear that $V = \overline{W}_0 \cup W_1 \cup \overline{W}_1 \cup \ldots, \cup \overline{W}_{n-1} \cup W_n \cup \ldots$ is PL homomorphic to $M' \times [0, \infty)$. Hence $M \times [0, \infty) = W_0 \cup V$ is PL homomorphic to $W \cup (M' \times [0,1))$. But, using the fact that M' is PL collared in W, it follows that $W \cup (M' \times [0,1)) \approx W - M'$ and the result follows.

In concluding this section we will describe the construction of Siebenmann and Sondow of a sequence $\{K_i\}_{i=0}^{\infty}$ of pwl $(n$-$2)$-spheres in $S^n (n \geqslant 6)$ such that the pairs $\{(S^n, K_i)\}_{i=0}^{\infty}$ are all topologically equivalent, but no two are pwl homomorphic.

The $(n$-$2)$-sphere $K_0 \subset S^n$ is constructed by first appropriately twist-spinning the trefoil knot to obtain a locally flat pwl $(n$-$3)$-sphere $\Sigma \subset S^{n-1}$, and then suspending Σ to get $K_0 \subset S^n$. ($\Sigma \subset S^{n-1}$ is locally flat if for each $p \in \Sigma$ there exists a neighborhood N of p in S^{n-1} such that the pair $(N, N \cap \Sigma)$ is topologically equivalent to the pair $(E^{n-1}, E^{n-3}\times(0,0))$.) We note that K_0 (and each of the other K_i) will be locally knotted at precisely two points, namely the two suspension points (that is, if v is one of the suspension points of $K_0\subset S^n$ and α is any subdivision of the pair (S^n,K_0), then the pair $(\mathrm{lk}(v,\alpha S^n), \mathrm{lk}(v,\alpha K_0))$ is not topologically equivalent to the standard unit $(n-1,n-3)$-sphere pair $(S^{n-1}\ S^{n-3})$).

We first describe briefly the spinning construction. Let J be a pwl $(n-3)$-sphere in $E^{n-1}_+ = \{x = (x_1,\ldots,x_{n-1},x_n)\in E^n \mid x_{n-1}\geqslant 0, x_n = 0\}$ which intersects $E^{n-2} = \mathrm{Bd}\,E^{n-1}_+$ in an $(n-3)$-simplex Δ, and let D be the closure of J-Δ. Let $\alpha_t\colon E^{n-1}_+ \longrightarrow E^n$ be the rigid rotation in

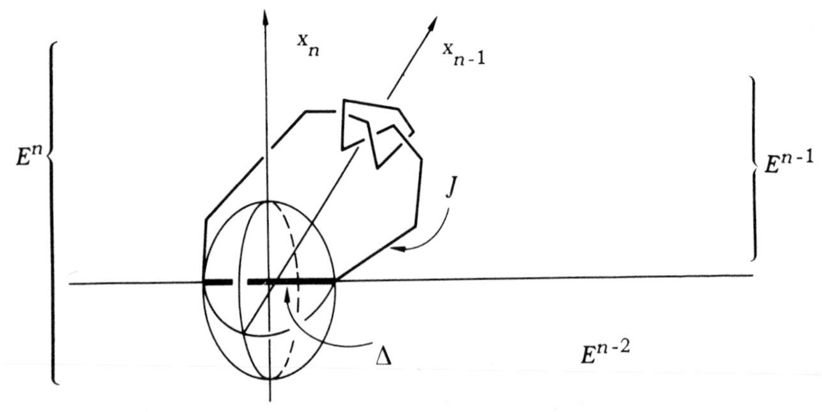

$E^n = (y_1,\ldots,y_n)$ of $E^{n-1}_+ = (x_1,\ldots,x_{n-1})$ defined by the equations

$$y_i = x_i \quad i\leqslant n-2$$

$$y_{n-1} = x_{n-1}\cos t$$

$$y_n = x_{n-1}\sin t.$$

Then the set $K = \{\alpha_t(x)\in E^n \mid x\in D \text{ and } t\in[0,2\pi]\}$ is clearly an $(n-2)$-sphere in $E^n\subset S^n$ and is said to be the result of the spinning J about E^{n-2} in E^n.

By rotating paths in E^n-K into E^{n-1}-J, it is easily verified that $\pi_1(E^n\text{-}K) \cong \pi_1(E^{n-1}\text{-}J)$. For instance, if J is the trefoil knot (as indicated in the above figure) and $n = 4$, then we obtain a 2-sphere $K \subset E^4$ which is knotted because $\pi_1(E^4\text{-}K) \cong \pi_1(E^3\text{-}J) \ncong Z$, since the latter group is isomorphic to the group having a presentation $\{a, b \mid aba = bab\}$. (Refer to Appendix B for a brief discussion of knot theory and the fact that $\pi_1(E^3\text{-}J)$ is as described above.)

What we actually want to do, however, is to not only spin the arc D about E^2 in E^4, but at the same time to twist the 'knotted part' of the arc through 5 revolutions while it is completing one spin about E^2. We will now give a description of this '5-twist-spinning' of the trefoil knot. The construction described below can also be carried out in general for knots (S^n, Σ^{n-2}), where Σ^{n-2} is a locally flat $(n\text{-}2)$-sphere in the n-sphere S^n, to give a locally flat knot (S^{n+1}, Σ^{n-1}) of one dimension higher by a corresponding k-twist-spinning process, where k is an integer. For further details and also for the proofs of some of the results we will assume, the reader can refer to E.C. Zeeman's article, 'Twisting spun knots', *Transactions of the American Mathematical Society* 115 (1965), 471-495.

Let Δ^3 denote the unit ball in Euclidean 3-space E^3. We introduce latitude and longitude coordinates in the unit sphere Bd Δ^3 as follows. We write E^3 as the product $E^3 = E^1 \times E^2$. Given a point $z \in \text{Bd}\ \Delta^3$, define the latitude of z to be the projection of z on E^1 and the longitude of z to be the angular polar coordinate of the projection of z on Δ^2 in E^2. Therefore, the latitude of z is a unique point $x \in \Delta^1$ and the longitude of z is either a unique point ϕ of S^1 if $z \notin \text{Bd}\ \Delta^1$, or else is indeterminate if $z \in \text{Bd}\ \Delta^1$.

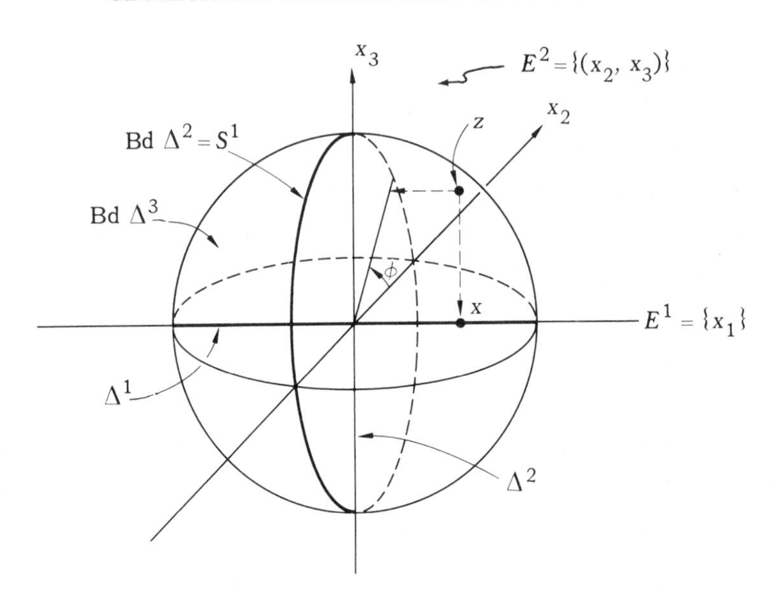

Let J be the trefoil knot in $E^3 \subset S^3$ as given above, so that $\Delta^3 \cap J = J \cap E^2 (=\{(x_1, x_2)\}) = \Delta^1 \subset J$ is the unit 1-ball, where Δ^3 is the unit 3-ball in E^3. Let (D^3, D^1) be the ball pair (D^3, D^1) = the closure of $[(S^3, J) - (\Delta^3, \Delta^1)]$. Then (D^3, D^1) is a knotted ball pair, in effect the same knot as the given knot (S^3, J). We now introduce latitude and longitude coordinates (x, ϕ) into Bd D^3 = Bd Δ^3 as above, where $x \in \Delta^1$ and $\phi \in S^1$ = Bd Δ^2. Let D^2 denote a copy of the unit 2-ball in E^2 and let (r, θ) be polar coordinates in D^2. We now consider (Bd $D^3 \times D^2$, Bd $D^1 \times D^2$) = X and ($D^3 \times$ Bd D^2, $D^2 \times$ Bd D^2) = Y. Let f: Bd $X \to$ Bd Y be the homomorphism given by $(x, \phi) \times (1, \theta) \to (x, \phi + 5\theta) \times (1, \theta)$ where $(x, \phi) \in$ Bd D^3 and $(1, \theta) \in$ Bd D^2. We define the 5-twist-spun knot to be the pair given by $X \cup_f Y$ = (Bd $D^3 \times D^2 \cup_f D^3 \times$ Bd D^2, Bd $D^1 \times D^2 \cup_f D^1 \times$ Bd D^2). It is clear that Bd $D^1 \times D^2 \cup_f D^1 \times$ Bd D^2 is a 2-sphere. To see that Bd $D^3 \times D^2 \cup_f D^3 \times$ Bd D^2 is a 4-sphere, we note that the homomorphism f of Bd $D^3 \times$ Bd D^2 onto itself given by f can be extended to a homomorphism, f' say, of $D^3 \times$ Bd D^2 *to*

itself (just twist the interior of the ball D^3 along with its boundary). Hence there is a homomorphism $\text{Bd}(D^3 \times D^2) = \text{Bd}\, D^3 \times D^2 \cup_1 D^3 \times \text{Bd}\, D^2 \xrightarrow{1 \cup f'} \text{Bd}\, D^3 \times D^2 \cup_f D^3 \times \text{Bd}\, D^2$.

Let us denote by K' the above 2-sphere in S^4 obtained by 5-twist-spinning the trefoilf knot. It turns out that $\pi_1(S^4\text{-}K') \cong G \times Z$, where G is the binary icosahedral (dodecahedral), group. G has the presentation $G = \{a, b | a^3 = b^5 = (ab)^2\}$. It can be proved that $z = a^3 = b^5 = (ab)^2$ is an element of order 2 in G, hence $G/\{1, z\} \cong \{a, b | a^3 = b^5 = (ab)^2 = 1\}$ is the familiar alternating group A_5 of even permutations on 5 symbols, which is a group of order 60. (Therefore G is a finite group of order 120). G is called the binary icosahedral (dodecahedral) group because A_5 is also the group of rotations of a regular icosahedron (dodecahedron), which has 20 triangular faces, 5 of which meet at each vertex (12 pentagonal faces, 3 of which meet at each vertex). By considering the elements of order 5 of A_5 which correspond to rotating the icosahedron (dodecahedron) through an angle of $2\pi/5$ about an axis determined by a vertex (the center of a face) and the center of the solid, and lifting these elements into G, we obtain the 5-sylow subgroups of $G \times Z$. We state the following algebraic lemma without proof.

LEMMA I.40. *If $\phi: T_5 \to G \times Z$ is the inclusion of a 5-sylow subgroup, then $\phi_* \text{Wh}(T_5) \subset \text{Wh}(G \times Z)$ is infinite cyclic (recall that* $\text{Wh}(T_5)$ *is infinite cyclic). Also if θ is any automorphism of $G \times Z$, then θ_* maps $\phi_* \text{Wh}(T_5)$ onto itself.*

Now by spinning the above 2-sphere K' in E^4 (without twisting), we obtain a 3-sphere in E^5 whose complement has fundamental group $G \times Z$. By iteration, and making the resulting spheres pwl, we obtain the following.

LEMMA I.41 (Zeeman). *For each $n \geqslant 4$, there exists a pwl $(n-2)$-sphere $K \subset S^n$ such that $\pi_1(S^n\text{-}K) \cong G \times Z$ (where G is the binary icosahedral group as above).*

The counterexamples to the Hauptvermutung for sphere pairs result from the construction of non-trivial invertible h-cobordisms on the sphere pair (S^n,K) of Lemma I.41, in essentially the same way that counterexamples to the Hauptvermutung for polyhedra result from the construction of non-trivial invertible h-cobordisms on a closed PL-manifold. We now give some details.

By a proper annulus pair is meant a PL-manifold pair (W^{n+1}, M^{n-1}) such that

(a) $W = S^n \times I$

(b) $M \approx S^{n-2} \times I$

(c) $M \cap \text{Bd } W = \text{Bd } M$

(d) $M \cap (S^n \times \{i\})$ is a PL $(n$-$2)$-sphere $K_i \times \{i\}$, $i = 0,1$.

We identify (S^n, K_i) naturally with $(S^n \times \{i\}, K_i \times \{i\})$.

(e) the regular neighborhood N of M in W is a PL product, $N \approx M \times I^2$.

Given a proper annulus pair (W, M), we say that the triple $C = \{(W, M);\ (S^n, K_0), (S^n, K_1)\}$ is a strong knot h-cobordism from the knot (S^n, K_0) to the knot (S^n, K_1) if and only if the inclusion $S^n\text{-}K_i \subset W\text{-}M$ is a homotopy equivalence, $i = 0,1$.

We define the torsion of this strong knot h-cobordism by $\tau(C) = \tau(W\text{-}\mathring{N}, (S^n \times \{0\}\text{-}\mathring{N}) \in \text{Wh}(\pi_1(S^n\text{-}K_0))$ where $\mathring{N}$ is an open regular neighborhood of M in W and we have identified the groups $\text{Wh}(\pi_1(S^n\text{-}K_0))$ and $\text{Wh}(\pi_1(S^n \times \{0\}\text{-}\mathring{N}))$ by the inclusion-induced isomorphism.

In order to work with strong knot h-cobordisms, we will need the s-cobordism theorem and Stallings's existence theorem for relative h-cobordisms. A relative h-cobordism is an ordered triple $C = (W;M,M')$ such that

(a) W, M, and M' are PL-manifolds

(b) M and M' are disjoint submanifolds of Bd W

(c) M and M' are both deformation retracts of W,

(d) C is a product along its boundary, that is

Bd W-int M-int $M' \approx (\text{Bd } M) \times I$

We define $\tau(C) = \tau(W,M) \in \text{Wh}(\pi_1(M))$ and the s-cobordism and existence theorem, previously stated for 'absolute' h-cobordisms, then extend to hold also for relative h-cobordisms (again under the added hypothesis that the manifolds involved admit compatible differentiable structures). By passing always from the strong knot h-cobordism $C = \{(W,M); (S^n, K_0), (S^n, K_1)\}$ to the relative h-cobordism $D = \{W\text{-}\mathring{N}; (S^n \times \{0\})\text{-}\mathring{N}, (S^n \times \{1\})\text{-}\mathring{N}\}$ the following facts can be established:

(i) $\tau(C) = 0$ if and only if (W^n, M) is trivial $(n \geqslant 6)$, that is $(W^n, M) \approx (S^{n-1} \times I, K_0 \times I)$;

(ii) the torsion of the dual $\overline{C}$ of C, $\overline{C} = \{(W,M); (S^n, K_1), (S^n, K_0)\}$ obtained by interchanging the ends of C, is

$\tau(\overline{C}) = (-1)^n \overline{\tau(C)}$ where the bar over $\tau(C)$ denotes the involution of $\text{Wh}(\pi_1(S^n\text{-}K_0))$ induced by the involution $\sigma \to \sigma^{-1}$ of $\pi_1(S^n\text{-}K_0)$;

(iii) if the right end of C is 'glued' to the left end of $C' = \{(W',M'); (S^n, K_0'), (S^n, K_1')\}$ so that $(W \cup W', M \cup M')$ gives a strong knot h-cobordism $C \cup C'$ from (S^n, K_0) to (S^n, K_1') then $\tau(C \cup C') = \tau(C) + \tau(C')$; and

(iv) given a knot (S^n, K_0) with $n \geqslant 5$ and $\tau_0 \in \text{Wh}(\pi_1(S^n\text{-}K_0))$, there exists a strong knot h-cobordism C with (S^n, K_0) as left end and $\tau(C) = \tau_0$.

We note that (i), (iii) and (iv) imply, just as in the proof for the absolute case, that every strong knot h-cobordism of dimension $\geqslant 6$ is invertible (in the obvious sense).

We are now ready to construct the examples. Given $n \geqslant 6$, let (S^{n-1}, L^{n-3}) be a PL knot such that $\pi_1(S^{n-1}\text{-}L^{n-3}) \cong G \times Z$ as provided by Lemma I.41. Let β be a generator of the infinite cyclic subgroup $\phi_* \mathrm{Wh}(T_5) \subset \mathrm{Wh}(\pi_1(S^{n-1}\text{-}L))$ of Lemma I.40. Then, for each $k = 0, 1, 2, \ldots$ let $C_k = \{(W_k^n, M_k); (S^{n-1}, L^{n-3}), (S^{n-1}, L_k)\}$ be an invertible strong knot h-cobordism with $\tau(C_k) = k\beta$ as given by (iv) above.

LEMMA I.42. *If $i \neq j$, there does not exist a* PL *homomorphism* $f\colon (W_i, M_i) \to (W_j, M_j)$.

Proof. For if f were such a PL homomorphism and θ is the automorphism of $\pi_1(S^{n-1}\text{-}L)$ induced by $f|_{S^{n-1}\text{-}L}$, then $\theta_* \tau(C_i)$ would be either $\tau(C_j)$ or $\tau(\overline{C}_j)$, depending upon whether f maps the left end of C_i to the left or right end of C_j. But $\theta_* \tau(C_i) = \pm i\beta$ is not equal to either $\tau(C_j) = j\beta$ or $\tau(\overline{C}_j) = \pm \overline{j\beta} = +(\pm j\beta)$.

We now construct the PL knot (S^n, K_k) by adding a cone over each end of the strong knot h-cobordism (W_k, M_k), $k \geqslant 0$.

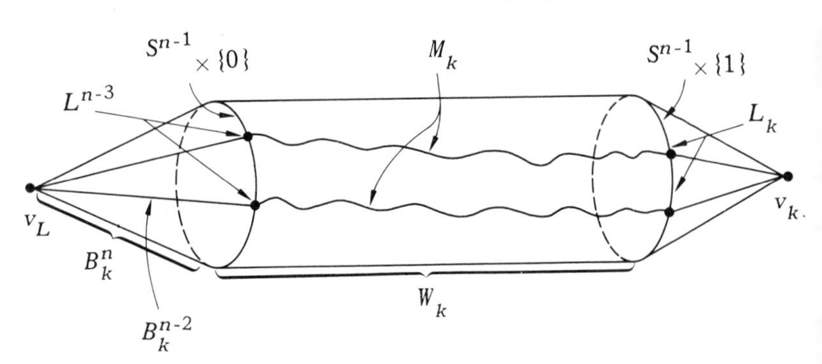

We note that the PL $(n\text{-}2)$-sphere K_k is locally knotted precisely at the two vertex points. These knots $\{(S^n, K_k)\}_{k=0}^{\infty}$ form our collection of examples in dimension $n \geqslant 6$.

THEOREM I.43. *If $i \neq j$, then (S^n, K_i) and (S^n, K_j) are not* PL *homomorphic.*

Proof. Suppose $g: \alpha(S^n, K_i) \to \beta(S^n, K_j)$ is a simplicial homomorphism. Then g must carry the two locally knotted points to K_i to the two locally knotted points of K_j. If we suppose that α and β are fine enough subdivisions so that the stars of these locally knotted points miss the corresponding ends of (W_i, M_i) or (W_j, M_j), then if we excise the open stars of all these locally knotted points, it follows, from a relative version of Lemma I.37, that what remains is a PL homomorphism between a copy of (W_i, M_i) and a copy of (W_j, M_j). But this contradicts Lemma I.42.

THEOREM I.44. *If the strong knot h-cobordism $C = \{(W, M);$ $(S^n, K_0), (S^n, K_1)\}$ is invertible, then $(W,M) - (S^n, K_1) \approx (S^n, K_0) \times [0, 1)$.*

Proof. The proof of this theorem is essentially the same as that of the corresponding result in the absolute case (Proposition I). If C' is an inverse to C, then we have formally that

$$\begin{aligned}(S^n,K_0) \times [0,1) &\approx (S^n,K_0) \times [0,\infty) = \bigcup_{i=1}^{\infty} ((S^n,K_0) \times [i\text{-}1, i]) \\ &\approx (C \cup C') \cup (C \cup C') \cup \ldots \approx C \cup [(C' \cup C) \cup (C' \cup C) \cup \ldots] \\ &\approx (W,M) \cup ((S^n,K_1) \times [0,\infty)) \approx (W,M) \cup ((S^n,K_1) \times [0,1)) \\ &\approx (W,M) - (S^n, K_1).\end{aligned}$$

It now follows easily from Theorem I.44 that the knots (S^n, K_k) are all topologically equivalent. For let (B^n_k, B^{n-2}_k) denote the cone-pair over the left boundary pair (S^{n-1}, L^{n-3}) of (W_k, M_k) and let v_k be the vertex of the cone on the right boundary pair (S^{n-1}, L_k). Then $(S^n, K_k) - \text{int}\,(B^n_k, B^{n-2}_k) - \{v_k\}$ is PL homomorphic to $(S^n, K_0) - \text{int}\,(B^n_0, B^{n-2}_0) - \{v_0\}$ because both are these are PL homomorphic to $(S^{n-1}, L) \times [0,1)$ by Theorem I.44. Hence there is a homomorphism f between their 1-point compactifications $(S^n, K_k) - \text{int}\,(B^n_k, B^{n-2}_k)$ and $(S^n, K_0) - \text{int}\,(B^n_0, B^{n-2}_0)$. Extending f so as to take (B^n_k, B^{n-2}_k) onto (B^n_0, B^{n-2}_0), we obtain a homomorphism of (S^n, K_k) onto (S^n, K_0).

CHAPTER II

Complexes in Euclidean Space

§ A. Unknotting Combinatorial Balls

The following section is basically that given in the paper by E.C. Zeeman, 'Unknotting Combinatorial Balls', *Annals of Math.* 78 (1963), 501–526. Here, also we will use polyhedra (i.e., a topological space underlying a finite simplicial complex) and by a subpolyhedron we will mean the subspace underlying a subcomplex of some rectilinear subdivision. All manifolds and all maps or homomorphisms are pwl unless otherwise stated. A pwl embedding of a q-ball B^q in a p-ball B^p is called proper if $\dot{B}^q \subset \dot{B}^p$ and int $B^q \subset$ int B^p. We call p-q the codimension.

We define a (p, q)-sphere pair (S^p, S^q), $p > q$, to be a pair of spheres such that S^q is a subcomplex of S^p. We define a (p, q)-ball pair (B^p, B^q), $p > q$, to be a pair of balls such that B^q is a subcomplex of B^p and B^q is properly embedded in B^p. We will denote the boundary sphere pair of a ball pair $Q = (B^p, B^q)$ by $\dot{Q} = (\dot{B}^p, \dot{B}^q)$. The join of a sphere pair (S^p, S^q) to a sphere S^r is the sphere-pair $(S^p S^r, S^q S^r)$. Similarly, the join of a sphere-pair to a ball, or the join of a ball-pair to a sphere or a ball, gives a ball-pair. In particular, the join of a pair to a point is called a cone-pair, and is an example of a ball pair.

If $X = (X^p, X^q)$, $Y = (Y^r, Y^s)$ are two pairs we say X contains Y if Y^r is a subcomplex of X^p and $Y^s = X^q \cap Y^r$. In particular, if

$(S^p, S^q) \supset (B^p, B^q)$ then $(S^p - \mathring{B}^p, S^q - \mathring{B}^q)$ is a ball pair. Also, if two pairs $\underset{X}{\underset{\parallel}{(X^p, X^q)}}$, $\underset{Y}{\underset{\parallel}{(Y^p, Y^q)}}$ are contained as subcomplexes of a larger complex, then we can define $X \cup Y = (X^p \cup Y^p, X^q \cup Y^q)$ and $X \cap Y = (X^p \cap Y^p, X^q \cap Y^q)$. Two pairs (X^p, X^q), (Y^p, Y^q) are said to be homomorphic if there is a pwl homomorphism carrying X^p onto Y^p which also carries X^q onto Y^q. If K is a complex, let ΣK denote the suspension of K. The nth suspension is defined inductively, $\Sigma^n K = \Sigma(\Sigma^{n-1} K)$, $\Sigma^1 K = \Sigma K$. We define the standard (p, q)-ball to be $\Gamma^{p,q} = (\Sigma^{p-q} \Delta^q, \Delta^q)$ and the standard (p, q)-sphere pair to be the boundary $\dot{\Gamma}^{p+1, q+1}$ of the standard $(p\ 1, q\ 1)$-ball pair.

We define a ball pair or a sphere pair to be unknotted if it is homomorphic to a standard pair. In general, if S^q is pwl embedded in S^p, rather than being a subcomplex of S^p, we say that S^q is unknotted in S^p, if, having chosen a subdivision S_1^p of S^p that contains a subcomplex S_1^q covering the image of the embedded S^q, the sphere pair (S_1^p, S_1^q) is unknotted. Similarly for balls.

THEOREM II.1. *If p-$q \geqslant 3$, then any (p, q)-ball pair is unknotted.*

THEOREM II.2. *If p-$q \geqslant 3$, then any (p, q)-sphere pair is unknotted.*

The proof will be by induction on p, keeping the codimension fixed. Let $\mathcal{B}(p, q)$ be the statement that Theorem II.1 is true for any (p, q)-ball pair and $\mathcal{S}(p, q)$ the statement that Theorem II.2 is true for any (p, q)-sphere pair. We will first show that $\mathcal{B}(p, q)$ implies $\mathcal{S}(p, q)$ and then with the help of ten lemmas, that $\mathcal{B}(p\text{-}1, q\text{-}1)$ and $\mathcal{S}(p\text{-}1, q\text{-}1)$ imply $\mathcal{B}(p, q)$, provided $p \geqslant q+3$. For codimension r, the induction begins trivially with $\mathcal{B}(r, 0)$. Since the inductive steps are the same for all codimensions $\geqslant 3$, we can drop the suffix q, and let $\mathcal{B}(p)$ denote the inductive assumption that Theorem II.1 is true for all (p', q)-ball pairs such that $p \geqslant p' \geqslant q+3$. Similarly we use $\mathcal{S}(p)$ for Theorem II.2.

Remark. Theorem II.2 implies that, provided the codimension is $\geqslant 3$, spheres unknot in Euclidean space. That is, given a pwl embedding of S^q in E^p, $p-q \geqslant 3$, then there exists a pwl homomorphism of E^p onto itself throwing S^q onto the boundary of a $(q+1)$-simplex. To see this, embed E^p in S^p piecewise linearly onto the complement of a point x, say. By Theorem II.2, there is a homomorphism $h: (S^p, S^q) \longrightarrow \dot{\Gamma}^{p+1, q+1}$ onto the standard sphere pair. We now pick a suspension vertex y such that $h(x) \notin y\dot{\Delta}^{q+1}$. Then $h^{-1}(y\dot{\Delta}^{q+1})$ gives a $(q+1)$-ball in E^p spanning S^q. By results of Section D, Chapter IV, of Volume I, we can throw this onto a $(q+1)$-simplex by a homomorphism of E^p.

LEMMA II.3. $\mathcal{B}(p) \Rightarrow \mathcal{S}(p)$.

Proof. Let $p = (S^p, S^q)$ be a sphere pair with $p-q \geqslant 3$ and pick a vertex $x \in S^q$. Let Q be the ball-pair given by $(S^p - \overset{\circ}{\mathrm{st}}(x, S^p), S^q - \overset{\circ}{\mathrm{st}}(x, S^q))$. Then $P = Q \cup x\dot{Q}$. Let y be the vertex of Δ^{q+1} opposite the face Δ^q. Then $\dot{\Gamma}^{p+1, q+1} \cong \Gamma^{p,q} \cup y\dot{\Gamma}^{p,q}$.

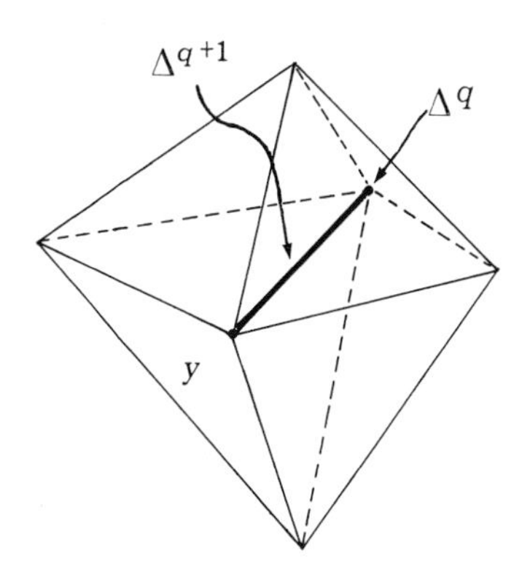

Any homomorphism $Q \longrightarrow \Gamma^{p,q}$ given by $\mathfrak{B}(p)$ can be extended by mapping $x \longrightarrow y$ to a homomorphism $P \longrightarrow \dot{\Gamma}^{p+1,q+1}$. Hence P is unknotted. The same argument holds for all $p' \leqslant p$.

LEMMA II.4. *Let Q_1, Q_2 be two unknotted (p,q)-ball pairs. Then any homomorphism $h: \dot{Q}_1 \longrightarrow \dot{Q}_2$ can be extended to a homomorphism $Q_1 \longrightarrow Q_2$.*

Proof. Let Γ be the standard (p,q)-ball pair, $\Gamma = (\Sigma^{p-q}\Delta^q, \Delta^q)$ and let y be the barycenter of Δ^q. For $i = 1, 2$, since the pair Q_i is unknotted there exists a homomorphism $Q_i \longrightarrow \Gamma$. Let f_i denote the composition $Q_i \longrightarrow \Gamma \longrightarrow y\dot{\Gamma}$. Let g denote the composition of the homomorphisms $\dot{\Gamma} \xrightarrow{(f_1|\dot{Q}_1)^{-1}} \dot{Q}_1 \xrightarrow{h} \dot{Q}_2 \xrightarrow{(f_2|\dot{Q}_2)} \dot{\Gamma}$.
Extend g to $g: y\dot{\Gamma} \longrightarrow y\dot{\Gamma}$. The composition
$Q_1 \xrightarrow{f_1} y\dot{\Gamma} \xrightarrow{g} y\dot{\Gamma} \xrightarrow{f_2^{-1}} Q_2$ is the homomorphism required to prove the lemma.

LEMMA II.5. *Assume $\mathfrak{B}(p\text{-}1)$, and suppose $p\text{-}q \geqslant 3$. Then if two unknotted $(p\text{-}q)$-ball pairs intersect in a common $(p\text{-}1, q\text{-}1)$-ball pair contained in the boundary sphere pair of each, then their union is also unknotted.*

Proof. Let $Q_1 \cap Q_2 = Q_3$, where $Q_3 \subset \dot{Q}_1$ and $Q_3 \subset \dot{Q}_2$. Let $Q_4 = \dot{Q}_1 - \mathring{Q}_3$ and $Q_5 = \dot{Q}_2 - \mathring{Q}_3$. Then by $\mathfrak{B}(p\text{-}1)$, Q_3, Q_4, and Q_5 are unknotted $(p\text{-}1, q\text{-}1)$-ball pairs and $\dot{Q}_3 = \dot{Q}_4 = \dot{Q}_5$. Let $\Sigma\Gamma$ be the suspension of the standard $(p\text{-}1, q\text{-}1)$-ball pair Γ, and let x_1, x_2 be the two suspension points. Choose a homomorphism $h: Q_3 \longrightarrow \Gamma$. By Lemma II.4 extend $h|_{\dot{Q}_4}$ to a homomorphism $Q_4 \longrightarrow x_1\dot{\Gamma}$ and extend $h|_{\dot{Q}_5}$ to a homomorphism $Q_5 \longrightarrow x_2\dot{\Gamma}$. Again by Lemma II.4, we can extend each of these to homomorphisms $Q_1 \longrightarrow x_1\Gamma$ and

$Q_2 \longrightarrow x_2\Gamma$ respectively. These two homomorphisms combine to give a homomorphism $Q_1 \cup Q_2 \longrightarrow \Sigma\Gamma$, which is clearly unknotted.

In some of our lemmas it will be convenient to use two types of collapsings, namely simplicial collapsings (defined as in earlier sections) and polyhedral collapsings, which we will now define. Let X be a polyhedron and Y a subpolyhedron. We say there is an elementary P-collapse from X to Y if there exist complexes K, L triangulating X, Y and a ball B^n with a face B^{n-1}, such that $K = L \cup B^n$ and $B^{n-1} = L \cap B^n$. We say X P-collapses to Y, written $X \overset{P}{\searrow} Y$, if there exists a finite sequence of elementary P-collapses going from X to Y. If X P-collapses to a point, we call X P-collapsible, and write $X \overset{P}{\searrow} 0$.

LEMMA II.6. *If L is a subcomplex of K, then $K \overset{P}{\searrow} L$ if and only if there is a subdivision αK of K such that $\alpha K \searrow \alpha L$* (refer to Theorem III.6 in Volume I).

Definition. Let M be an n-manifold and X a subpolyhedron. A polyhedral regular neighborhood N of X in M is a subpolyhedron of M such that:

(i) N is a closed neighborhood of X in M,

(ii) N is an n-manifold, and

(iii) $N \overset{P}{\searrow} X$.

LEMMA II.7. *If N_1, N_2 are two polyhedral regular neighborhoods of X in M, then there is a homomorphism $N_1 \longrightarrow N_2$ keeping X fixed* (refer to Section B, Chapter III in Volume I).

LEMMA II.8. *Assume $\mathcal{B}(p-1)$ and $\mathcal{S}(p-1)$. If (B^p, B^q), $p-q \geqslant 3$ is a ball pair such that $B^p \overset{P}{\searrow} B^q$, then it is unknotted.*

Proof. A ball is P-collapsible; hence, choose a subdivision αB^q of B^q so that $\alpha B^q \searrow 0$. Extend this to a subdivision αB^p of B^p. Let βB^p be the second derived complex of αB^p. Let N be the closed simplicial neighborhood of βB^q in βB^p. Now by hypothesis, B^p itself is also a regular neighborhood of B^q in B^p. Therefore Lemma II.7 gives a homomorphism between the ball-pairs $(B^p, B^q) \longrightarrow (N, \beta B^q)$. Therefore to prove the lemma, it suffices to show that $(N, \beta B^q)$ is unknotted. Let $\alpha B^q = K_k \searrow K_{k-1} \searrow \ldots \searrow K_1 \searrow K_0 = x$ be the simplicial collapse of αB^q down to a point x. Let Q_i be the ball pair consisting of the closed simplicial neighborhoods of K_i in $(\beta B^p, \beta B^q)$. We shall show inductively that Q_i is unknotted.

The induction begins with $Q_0 = xL_x$, the cone pair on $L_x = (\mathrm{lk}\,(x, \beta B^p), \mathrm{lk}\,(x, \beta B^q))$. Now L_x is either a sphere pair if x is in the interior of αB^q, or a ball pair if x is on the boundary of αB^q, but in either case is unknotted by hypothesis, $\mathcal{B}(p\text{-}1)$ and $\mathcal{S}(p\text{-}1)$. Hence Q_0 is unknotted.

For the inductive step, suppose Q_{i-1} is unknotted, where $1 \leqslant i \leqslant k$. Since $K_i \searrow K_{i-1}$ is an elementary simplicial collapse, $K_i - K_{i-1}$ consists of a simplex A with a free top-dimensional face C, say. Let a, c denote the barycenters of A, C, respectively. Let $Q_a = aL_a$, the cone-pair on $L_a = (\mathrm{lk}\,(a, \beta B^p), \mathrm{lk}\,(a, \beta B^q))$, which is unknotted for the same reason as Q_0 above. Similarly let $Q_c = cL_c$, which is also unknotted. Now $Q_i = Q_{i-1} \cup Q_a \cup Q_c$. But Q_{i-1} and Q_a intersect in a common $(p\text{-}1, q\text{-}1)$ ball pair, and so do $(Q_{i-1} \cup Q_a)$ and Q_c. Hence applying Lemma II.5 twice, we see that Q_i is unknotted. At the end of the induction we have $Q_k = (N, \beta B^q)$ unknotted, which completes the proof of the lemma.

Remark. With codimension 2, it is possible to have $B^p \searrow B^{p-2}$, but (B^p, B^{p-2}) knotted. For example let (B^4, B^2) be the cone pair on a knotted (S^3, S^1). Then $B^4 \searrow B^2$ because a cone collapses onto any subcone. Also with codimension 2, it is possible to have a ball pair (B^p, B^{p-2}) such that B^p does not $\searrow B^{p-2}$, as for example a knotted arc in a 3-ball. The next lemma shows this cannot happen with higher codimension.

LEMMA II.9. *If (B^p, B^q), $p-q \geqslant 3$, is a ball pair, then $B^p \searrow B^q$.*

Once we have proved Lemma II.9 we are done; that is Lemmas II.8 and II.9 together provide the inductive step that $\mathcal{B}(p-1)$ and $\mathcal{S}(p-1)$ imply $\mathcal{B}(p)$. We shall postpone the proof of Lemma II.9 until after we prove some additional lemmas making use of some geometrical constructions we can get because of the codimension 3. Up to this point we have not had to work very hard, now things become a little sticky.

We now define the notion of shadows. Let I^p be the p-cube. We single out the last coordinate for special reference and write $I^p = I^{p-1} \times I$. We will think of I^{p-1} as horizontal and I as vertical, and we identify I^{p-1} with $I^{p-1} \times \{0\}$, the base of the cube. Let X be a polyhedron in I^p, a point of I^p is said to lie in the shadow of X if it lies vertically below some point of X. Let X^* be the closure of the set of points of X that lie in the same vertical line as some other point of X (i.e., the set of points of X that either lie in the shadow of X or else overshadow some other point of X). X^* is a subpolyhedron of X.

LEMMA II.10. *Let X be a polyhedron in I^p such that* $\dim = q < p$ *and* $\dim (X \cap I^p) \leqslant q-1$. *Then there exists a homomorphism of I^p onto itself throwing X into a position that satisfies the properties:*

(i) *X does not meet the top or the bottom of the cube,*

(ii) *X meets any vertical line finitely; and*

(iii) $\dim X^* \leqslant 2q-p+1$.

Proof. Choose a face I^{p-2} of I^{p-1}, so that $I^{p-2}\times I$ is a vertical top-dimensional face of I^p. Since $\dot{I}^p \not\subset X$ there is a homomorphism of I^p onto itself throwing $X \cap \dot{I}^p$ into the interior of this vertical face, satisfying condition (i). Now triangulate I^p so that X is a subcomplex. Then shift all the vertices of this triangulation by arbitrarily small moves into general position, in such a way that any vertex in the interior of I^p remains in the interior, and any vertex in a face of I^p remains inside that face. If the moves are sufficiently small, the new positions of the vertices determine an isomorphic triangulation, and a homomorphism of I^p onto itself. The general position ensures that X is thrown onto a polyhedron with the desired properties.

Suppose we are given polyhedra $Y \subset X \subset I^p$. If there is an elementary P-collapse from X to Y, define this collapse to be sunny if no point of X-Y lies in the shadow of Y. We say there is a sunny collapse $X \underset{S}{\overset{P}{\searrow}} Y$ is there exists a finite sequence of elementary sunny P-collapses going from X to Y. If there is a sunny collapse $X \underset{S}{\overset{P}{\searrow}} 0$, then X is called sunny collapsible. Similarly, we can define sunny simplicial collapses between complexes in I^p.

LEMMA I.11 (Corollary to Lemma II.6). *X is a sunny collapsible if and only if some triangulation of X is sunny simplicially collapsible.*

Proof. The proof of the lemma follows from work done earlier because each elementary sunny P-collapse can be factored into a sequence of elementary simplicial collapses, each of which will be sunny.

LEMMA II.12. *Suppose* (I^p, X) *is homomorphic to a* (p, q)*-ball pair,* p-$q \geqslant 3$ *and suppose* X *satisfies the* 3 *properties of Lemma II.10. Then* X *is sunny collapsible.*

Remark. Lemma II.12 fails with codimension 2. The classical example of a knotted arc in I^3 gives a good intuitive feeling for the obstruction to a sunny collapse: looking down from above, it is possible to start collapsing away until we hit underpasses, which are in the shadow and so prevent any further progress.

Proof of Lemma II.12. The proof is long, by a complicated induction. Let $Y_0 = \dot{\Delta}^q$ and $Z_0 = X$. We shall construct inductively two descending sequences of subpolyhedra

$$Y_0 \supset Y_1 \supset \ldots \supset Y_i \supset \ldots \supset Y_q$$

$$Z_0 \supset Z_1 \supset \ldots \supset Z_i \supset \ldots Z_q$$

so that for each i, $0 \leqslant i \leqslant q$, we have a homomorphism $f_i \colon C_i \longrightarrow Z_i$ from C_i onto Z_i, where C_i is the cone on Y_i, such that the following four properties are satisfied:

(1) Y_i is everywhere $(q$-i-$1)$-dimensional,

(2) $\dim Z_i^* \leqslant q$-i-2

(3) $f_i^{-1}(Z_i^*)$ does not contain the vertex of the cone C_i, and meets each generator of the cone finitely;

and

(4) there exists a sunny collapse $Z_{i-1} \searrow Z_i$.

The induction begins with $Z_0 = X$ and finishes with Z_q being a point (Y_q being empty). Therefore, the lemma will follow from Property (4), because the sequence $X = Z_0 \searrow Z_1 \searrow \ldots \searrow Z_q$ shows that X is sunny collapsible.

We have defined $Y_0 = \dot{\Delta}^q$, Z_0 X. Therefore Property (1) is trivial, because $\dot{\Delta}^q$ is everywhere $(q\text{-}1)$-dimensional. Property (2) follows from Lemma II.10-(iii), because dim $X^* \leqslant 2q\text{-}p+1 \leqslant q\text{-}2$ (here we use the fact that $p\text{-}q \geqslant 3$). Property (4) is vacuous because Z_{-1} is not defined. There remains to define the homomorphism f_0 so as to satisfy Property (3).

Choose a homomorphism $f\colon\ \Delta^q \longrightarrow X$ onto the q-ball $X = Z_0$. Choose a vertex v in the interior of Δ^q in general position with respect to $f^{-1}(X^*)$. General position means that $v \notin f^{-1}(X^*)$, and that any straight line in Δ^q through v meets f (X^*) finitely. Subdividing Δ^q at v gives a complex isomorphic to the cone C_0 on $Y_0 = \dot{\Delta}^q$. Define $f_0 = f\colon C_0 \longrightarrow Z_0$.

Fix i, $1 \leqslant i \leqslant q$. Suppose we are given the polyhedra Y_{i-1}, Z_{i-1} and the homomorphism $f_{i-1}\colon C_{i-1} \longrightarrow Z_{i-1}$ satisfying the four inductive properties. We have to define subpolyhedra Y_i, Z_i and a homomorphism $f_i\colon C_i \longrightarrow Z_i$, and prove the four properties for them. For simplicity, let us drop the suffix i-1, but retain the suffix i. That is, we are given $Y = Y_{i-1}$, $Z = Z_{i-1}$, $C =$ the cone on Y, and $f\colon C \longrightarrow Z$; and we shall eventually define Y_i, Z_i and f_i.

Let v be the vertex of the cone C, and let $W = f^{-1}(Z^*)$. Then W is a subpolyhedron of C of dimension $\leqslant q\text{-}i\text{-}1$ (by Property (2)) that does not contain v and meets each generator of the cone finitely (by Property (3)). Let $\pi\colon W \longrightarrow Y$ be the map defined by projecting from the vertex v onto the base Y of the cone C. Now in general π is not pwl, but we can get the following.

LEMMA II.13. *There exist triangulations K, L of W, Y such that for each simplex $A \in K$, πA is a simplex of L of the same dimension.*

Proof. Choose some triangulation of K_0 of W. For each simplex $A_0 \in K_0$, πA_0 is a simplex contained in Y. The dimension of πA_0 is the same as that of A_0 because of the inductive Property (3). As A_0 runs over the simplexes in K_0, the set of image simplexes πA_0 may criss-cross each other in Y, but, nevertheless, it is possible to find a triangulation L of Y, such that every πA_0 is covered by a subcomplex of L. Lift these subcomplexes under π to form a subdivision K of K_0. Then K, L satisfy the requirements of the lemma.

Let Y_i be the polyhedron underlying the $(q\text{-}i\text{-}1)$-skeleton of L (triangulating Y). By the inductive Property (1), Y is everywhere $(q\text{-}i)$-dimensional. Therefore every top dimensional simplex of L is $(q\text{-}i)$-dimensional, and Y_i is everywhere $(q\text{-}i\text{-}1)$-dimensional. Hence Property (1) holds for Y_i.

The cone $C_i = vY_i$ is a subcone of C. However, it is no good trying to define $f_i = f|_{C_i}$, because then we should have to have $Z_i = fC_i \supset fW = Z^*$, and so Z_i^* would in general be of dimension $q\text{-}i\text{-}1$, which is too high for Property (2). Here we must arrange some device for collapsing away the top-dimensional shadows of Z^*.

The first thing to observe is that the triangulation K of W is in no way related to the embedding of $fW = Z^*$ in the cube I^p. The images in I^p of the simplexes of K may link around and overshadow each other in an unpredictable fashion. Our next task is to take a subdivision K' of K that remedies this confusion.

Let $g : K \longrightarrow I^{p-1}$ denote the following composition:

$$\underbrace{K \subset C \longrightarrow Z \subset X \subset I^p \longrightarrow I^{p-1}}_{g}$$

Since g is pwl, we can find subdivisions K', M of K, I^{p-1} such that $g : K' \longrightarrow M$ is simplicial. Recall that $\dim K = \dim W \leqslant q\text{-}i\text{-}1$. Let $A_1, A_2, \ldots, A_m$ be the

$(q\text{-}i\text{-}1)$-simplexes of K'. Each A_j is mapped non-degenerately by g, by Lemma II.10. For each pair A_j, A_k, $j \neq k$, the interiors $\mathring{A}_j, \mathring{A}_k$ are mapped disjointly by f, and are either mapped disjointly or identified by g. If $gA_j \neq gA_k$ then no point of $f\mathring{A}_j$ overshadows any point of $f\mathring{A}_k$ and vice versa. If $gA_j = gA_k$, then vertical projections establish a homomorphism between $f\mathring{A}_j$ and $f\mathring{A}_k$, so that either $f\mathring{A}_j$ overshadows $f\mathring{A}_k$ or vice versa. Consequently overshadowing induces a partial ordering between the A's, and we choose the ordering $A_1, A_2, \ldots, A_m$ to be compatible with this partial ordering. We state this in the form of a lemma:

LEMMA II.14. *All points of X that overshadow $f\mathring{A}_k$ are contained $\cup_{1 \leqslant j < k} f\mathring{A}_j$.*

The next step is to construct a little $(q-i+1)$-dimensional blister J_j about each A_j in the cone C. The blisters are the device that enables us to make a sunny collapse, and the fact that there is just sufficient room to construct them is an indication of why codimension 3 is a necessary and sufficient condition for unknotting.

Choose $\varepsilon > 0$ and sufficiently small (the criterion for sufficiently small will appear at the end of the construction). There are two cases depending on whether or not A_j happens to lie in Y.

Case (i) Suppose $A_j \subset Y$; then the blister will lie at the bottom of the cone. Let a_j be the barycenter of A_j. Let b_j be the point on the line va_j a distance ε from a_j. Since $\dim A_j = \dim K' = q-i-1$, A_j is contained in a $(q-i-1)$ simplex D_j of K. Since $A_j \subset Y$, we also have $D_j \subset Y$, and so by Lemma II.13, $D_j = \pi D_j \in L$. By the inductive Property (1), Y is everywhere $(q-i)$-dimensional, and so there is at least one $(q-i)$-simplex $E_j \in L$ having D_j as a face. Let a_j' be the point of the join of

a_j to the barycenter of E_j, a distance ε from a_j. Define

$$J_j = a_j a'_j b_j \dot{A}_j .$$

Case (ii) Suppose $A_j \not\subset Y$; then the blister will lie in the middle of the cone. Again let a_j be the barycenter of A_j; then $a_j \notin Y$. By the inductive Property (3) the generator of the cone va_j through a_j does not meet A_j again. Let b_j denote the pair of points on this generator a distance ε on either side of a_j. As before, A_j is contained in a $(q-i-1)$-simplex D_j of K, only this time $D_j \not\subset Y$. By Lemma II.13, $\pi D_j \in L$, and again we can choose a $(q-i)$-simplex $E_j \in L$ having πD_j as a face. Let a'_j be the point on the line joining a_j to the barycenter of vE_j a distance ε from a_j. Again define J_j by by the same formula $J_j = a_j a'_j b_j \dot{A}_j$. Finally choose ε sufficiently small so that all the blisters are well defined, and so that no two overlap more than necessary, i.e., $J_j \cap J_k = \dot{A}_j \cap \dot{A}_k$ for each pair j, k.

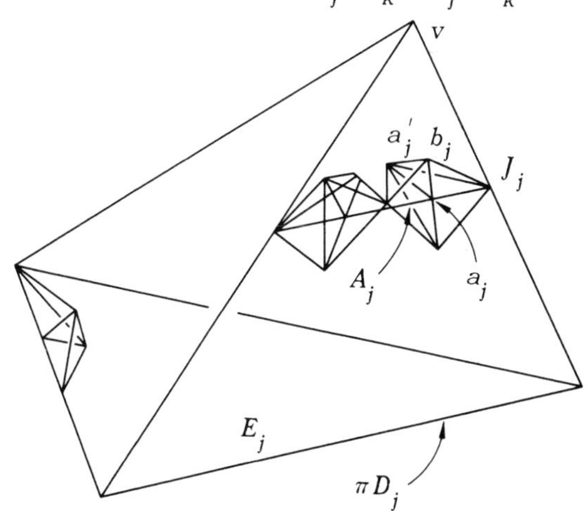

We are now in a position to complete the inductive definitions. Recall that we defined Y_i to be the $(q-i-1)$-skeleton of L. Hence $C_i = vY_i$, the cone on Y_i, is a subcone of C. Therefore, each j, $C_i \cap J_j = a_j b_j \dot{A}_j$. Define an embedding $e: C_i \longrightarrow C$ as follows:

let e be the identity outside all the blisters, and inside the jth blister map $a_j b_j \dot{A}_j$ linearly onto $a'_j b_j \dot{A}_j$, for each j. In other words to obtain the embedding $e: C_i \longrightarrow C$ from the inclusion $C_i \subset C$, we merely push up all the blisters. Define $f_i = fe: C_i \longrightarrow Z$ and define $Z_i = f_i C_i$. This completes the inductive definitions. We have already verified Property (1); there remains to verify properties (2), (3) and (4).

To verify Property (2), observe that $Z_i^* \subset Z_i \cap Z^* = fe C_i \cap fW = f(eC_i \cap W)$, since f is homomorphism. Hence $eC_i \cap W$ is contained in the $(q-i-2)$-skeleton of K'. Therefore $\dim Z_i^* \leqslant q-i-2$.

The same observation suffices to verify Property (3), because $f_i^{-1}(Z_i^*) \subset$ the $(q-i-2)$-skeleton of $K' \subset f^{-1}(Z^*)$. By induction, Property (3) holds for $f^{-1}(Z^*)$ and so it also holds for $f_i^{-1}(Z_i^*)$.

Finally we come to Property (4), which is the heart of the matter. Let $J = \cup_{1 \leqslant j \leqslant m} J_j$, the union of all the blisters. Given a $(q-i)$-simplex $E \in L$, let E_1, E_2 denote respectively the closures of $vE\text{-}J$, $E\text{-}J$. Now E_1 can be obtained from the $(q-i+1)$-simplex vE by removing one by one any blisters that happen to protrude into vE; therefore E_1, suitably triangulated, is a $(q-i+1)$-ball. Similarly E_2 is a $(q-i)$-ball, and a face of E_1. Removing the interiors of E_1 and E_2 defines an elementary collapse of C. Doing this successively for all the $(q-i)$-simplexes in L defines a collapse $C \searrow C_i \cup J$. But $C_i \cup J = eC_i \cup J$ and $f(eC_i \cup J) = Z_i \cup fJ$. Therefore the image under the homomorphism f of this collapse determines a sunny collapse $Z \searrow Z_i \cup fJ$, sunny because we have not yet removed any point of Z^*.

Now we collapse $eC_i \cup J \searrow eC_i$ by collapsing each blister in turn, $j = 1, 2, \ldots, m$, as follows. The blister J_j is a $(q-i+1)$-ball, and its intersection with eC_i (and all the other blisters) is the

$(q-i)$-face $a_j' b_j \dot{A}_j$. Therefore we may collapse J_j onto this face. The images under f of these collapses determine a sequence of elementary collapses $Z_i \cup (\cup_{j=1}^m fJ_j) \searrow Z_i \cup (\cup_{j=2}^m fJ) \searrow \ldots \searrow Z_i$. Each of these elementary collapses is sunny by Lemma II.14, and by virtue of our choice of the ordering $j = 1, 2, \ldots, m$; because, by the time we come to collapse fJ_k, say, the only points that might have been in shadows are those in the interior $f\mathring{A}_k$, but these are sunny for we have already removed everything that overshadows them. Hence we have demonstrated a sunny collapse.

$$Z \searrow Z_i \cup fJ \searrow Z_i.$$

We can now return to the proof of Lemma II.9, which will conclude the proofs of Theorem II.1 and II.2. We are given a ball-pair (B^p, B^q), $p - q \geqslant 3$, and we have to show that $B^p \searrow B^q$. By Lemmas II.10 and II.12, we can choose a homomorphism $B^p \longrightarrow I^p$ such that B^q is thrown onto a sunny collapsible polyhedron X satisfying the three properties of Lemma II.10. It suffices to show that $I^p \searrow X$.

Definition. If F is a complex or polyhedron in I^p, let $F^{\#}$ denote the polyhedron consisting of F together with all points of I^p lying in the shadow of F. Recall that I^{p-1} denotes the base of the cube. Let $M = I^{p-1} \cup X^{\#}$.

Proof that $I^p \searrow M$: The vertical projection $X \longrightarrow I^{p-1}$ is pwl and so we can choose triangulations of X, I^{p-1} with respect to which it is simplicial. Let L denote the triangulation of I^{p-1}. For each simplex $D \in L$, let $D \times I$ denote the prism lying vertically above D. If the interior of the prism meets X, then by Lemma II.10-(ii), it meets it in a finite number of simplexes, each of the same dimension as D and lying vertically above D. Let D_1 be the topmost of these; D_1 does not meet the top or bottom of the prism by Lemma II.10-(i). Then M contains the subprism bounded above by D_1 and contains no points above $\mathring{D}_1$. Let D' denote the subprism bounded below by D_1. If, on the other hand, the interior of $D \times I$ does not meet X, let $D' = D \times I$. Consider the elementary collapse of D' from the top onto the walls and base. Now enumerate the simplexes of L in order of decreasing dimension, and the corresponding sequence of elementary collapses determines a collapse $I^p \overset{P}{\searrow} M$. Next we make use of Lemma II.12 and Lemma II.11. Let K be a triangulation of X that is simplicially sunny collapsible by the sequence, say, of elementary simplicial sunny collapses

$$K = K_0 \searrow K_1 \searrow \ldots \searrow K_n = \text{a point}.$$

Let $M_i = I^{p-1} \cup X \cup K_i^{\#}$. In particular $M_0 = M$. We shall complete the proof of Lemma II.9 by showing that $I^p \searrow M_0 \searrow M_1 \searrow M_2 \searrow \ldots \searrow M_n \searrow X$. We already have the first step. The last step is easy, because $I^{p-1} \cap X = \emptyset$ by Lemma II.10-(i). Therefore M_n consists of I^{p-1} and X connected by a single arc $K_n^{\#}$. Therefore, collapse $M_n \searrow X$ by collapsing I^{p-1} onto the bottom point of the arc, and then collapsing the arc. Refer to the figures on the next page which illustrate the ideas.

There remain the intermediate steps $M_{i-1} \searrow M_i$, $1 \leqslant i \leqslant n$. We are given a sunny elementary simplicial collapse $K_{i-1} \searrow K_i$. Suppose that the collapse is across the simplex $A \in K_{i-1}$, from the face B. Let a be the vertex A opposite B. Therefore $K_i \cup A = K_{i-1}$ and $K_i \cap A = a\dot{B}$. Let a_1, A_1, B_1 be the vertical projections of a, A, B on the base I^{p-1} of the cube; A_1, B_1 are simplexes of the same dimension as A, B by Lemma 8 (ii). Let $A_1 \times I$ denote the prism lying above A_1. Let

$$U = (K_{i-1}^{\#} \cup I^{p-1}) \cap (A_1 \times I),\ V = (K_i^{\#} \cup I^{p-1}) \cap (A_1 \times I).$$

Then $M_{i-1} - M_i = U - (V \cup A)$. Therefore, to show $M_{i-1} \searrow M_i$ it suffices to prove that $U \searrow V \cup A$.

Let us examine U. U contains the subprism lying between A and A_1 (A does not meet A_1 by Lemma II.10–(i)). Since the collapse $K_{i-1} \searrow K_i$ is sunny, U contains no points above $\mathring{A} \cup \mathring{B}$. However U may contain material above $a\dot{B}$ in the walls $a_1 \dot{B}_1 \times I$ of the prism $A_1 \times I$.

Now examine V. To begin with V agrees with U in the walls $a_1 \dot{B}_1 \times I$. However, V does not contain A, and in face $A \cap V = a\dot{B}$. Since V is a polyhedron, we can find a point x vertically below the barycenter of B, such that $xA \cap V = a\dot{B}$. Let T be the closure of $U - xA$. Then $U \supset T \cup A \supset V \cup A$.

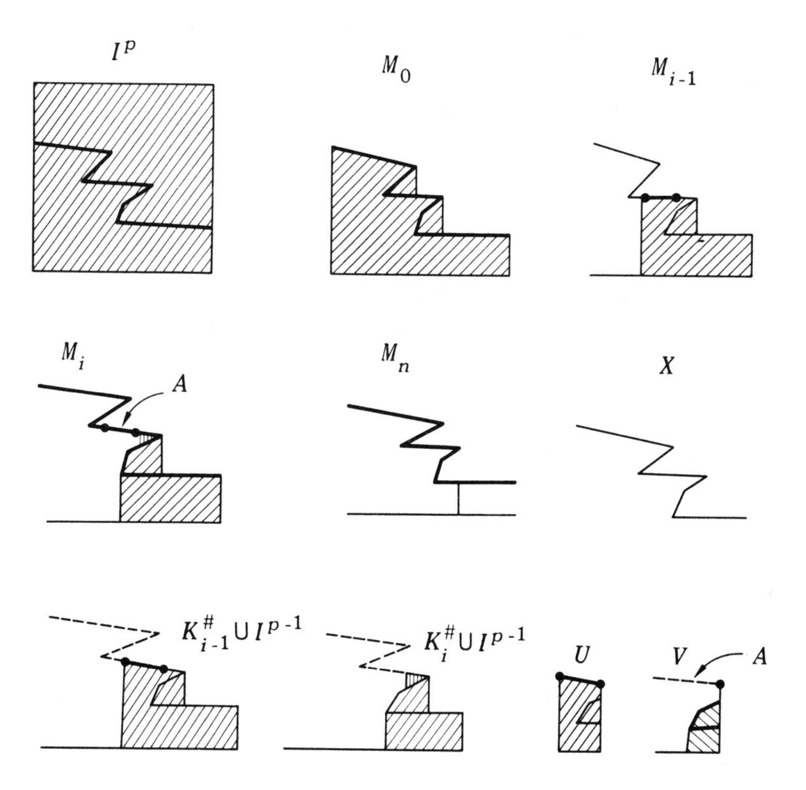

We shall show that we can collapse $U \searrow T \cup A \searrow V \cup A$. The first step is an elementary collapse across xA from the free face xB. For the second step, since $(T \cup A) - (V \cup A) = T - V$, it suffices to show that $T \searrow V$. We use the same device that we used at the beginning of the proof of this lemma. Choose triangulations of T, V, A_1 such that the vertical projections $T \longrightarrow A_1$ and $V \longrightarrow A_1$ are simplicial. Let L be the triangulation of A_1, and let L_0 denote the subcomplex covering $a_1 \dot{B}_1$.

Let D be a simplex of $L - L_0$, and let $E = D \cap L_0$ (possibly E is empty). The intersection of T with the prism $D \times I$ above D consists of a nondegenerate subprism bounded by a simplex, D_1 say, contained in $xa\dot{B}$, together with possibly some material above E. Similarly the intersection of V with $D \times I$ is a possibly degenerate subprism bounded above by a simplex, D_2 say, contained in $K_i \cup I^{p-1}$, together with the same material above E. Now $D_1 \neq D_2$ by our choice of x, and $D_1 \cap D_2$ is a common face, perhaps empty, above E. Therefore there is a nondegenerate subprism, D' say, bounded above by D_1 and below by D_2. Consider the elementary collapse of D' from the top D_1 onto the walls and base D_2. Now enumerate the simplexes of $L - L_0$ in order of decreasing dimension, and the corresponding sequence of elementary collapses determine the required collapse $T \searrow V$. Therefore we have shown $U \searrow V \cup A$, and hence $M_{i-1} \searrow M_i$, and hence $I^p \searrow X$.

§ B. Products with Intervals and Contractible Open Manifolds

By a psuedo n-cell we mean a contractible compact combinatorial n-manifold with boundary (whose boundary is not necessarily an $(n-1)$-sphere). Poenaru [52] (1960) and Mazur [39] (1961) gave the first examples of pseudo 5-cells which are not topological 4-cells and Curtis [17] (1961) gave examples for $n \geqslant 5$. Each of the examples failed to be topological n-cells because their boundaries were homology $(n-1)$-spheres that were not simply connected. These also have the property that when crossed with an interval, the cartesian product is homomorphic to an $(n+1)$-cell. By a result of Bing [3] such pathologies cannot happen for $n \leqslant 3$, i.e., if X is any topological space such that $X \times I = I^n$ $(n \leqslant 4)$ then X is homomorphic to I^{n-1}. Glaser [26] (1964) also gave further examples of pseudo n-cells $(n \geqslant 4)$ having the same properties with the additional results that each pseudo n-cell was the regular neighborhood of a contractible $(n-2)$-complex P^{n-2} embedded pwl in S^n so as to have a non-simply connected complement. Each P^{n-2} when crossed with an interval collapses to a point and hence their regular neighborhoods crossed with an interval were combinatorial $(n+1)$-cells, rather than just being topological $(n+1)$-cells. Also for $n \geqslant 5$ these regular neighborhoods can be expressed as the union of two combinatorial n-cells intersecting in a combinatorial n-cell, even though the boundary of the regular neighborhood itself is not simply connected. Glaser [27] (1965) constructed another pseudo 4-cell N^4 so that $N^4 \times I \approx I^5$, but in addition it is the union of two combinatorial 4-cells intersecting in a 4-cell. Making use of these examples one also obtains, for $n \geqslant 4$, examples of contractible open combinatorial n-manifolds which can be expressed as the union of two Euclidean n-spaces intersecting in a Euclidean n-space although the union itself is not homomorphic to E^n. By making use of the complement of a double Fox-Artin arc in S^3 one may also obtain an open contractible 3-dimensional example [27]. McMillan [40] (1962) constructed uncountably many nonhomomorphic open contractible 3-manifolds none of which can be embedded in E^3. Curtis and Kwun [20] (1965) constructed uncountably many non-homomorphic open contractible n-manifolds ($n \geqslant 5$) and Glaser [28] (1965) obtained uncountably many homomorphic open contractible 4-manifolds.

All of these contractible open n-manifolds when crossed with E^1 are homomorphic to E^{n+1}. For $n \geqslant 4$ this follows by Stallings's results (Section A, Chapter I), but in each case this fact may be obtained by merely making use of the particular constructions used. Also each of the examples (for $n \geqslant 4$) lie in E^n. In this section we will discuss some of these results in detail and indicate the main ideas of others. Also in this section, $=$ will usually mean topologically homomorphic.

LEMMA II.15. *If C^n is a compact n-manifold with boundary such that* int $C^n = E^n$, *and* $B =$ Bd $C^n = S^{n-1}$, *then* $C^n = I^n$.

Proof. By Brown's result [11] that the boundary of a manifold is collared (refer to Appendix A) there exists a homomorphism $h\colon B \times [0,1] \longrightarrow C^n$ such that $h(x,0) = x$ if $x \in B$. By the generalized Schoenflies theorem, the collared $(n-1)$-sphere $h(B \times \{1/2\})$ bounds a closed n-cell A in int $C^n = E^n$. Hence $C^n = A \cup h(B \times [0,1/2])$ is a closed n-cell.

PROPOSITION II.16. *If C^n is a compact contractible combinatorial n-manifold with boundary then:*

(i) *If $n \geqslant 5$ and* Bd $C^n = S^{n-1}$, *then C^n is homomorphic with I^n.*

(ii) *If $n \geqslant 6$ and* int $C^n = E^n$, *then $C^n = I^n$.*

Proof. (i) follows from Lemma II.15, after we show int $C^n \approx E^n$.

(ii) follows from Lemma II.15, after we show Bd $C^n = S^{n-1}$.

Proof of (i): Int $C^n \approx E^n$, since int C^n is a contractible open combinatorial manifold 1-connected at infinity (since Bd $C^n = S^{n-1}$ and Bd C^n is collared in C^n, it follows that int C^n is 1-connected at ∞).

Proof of (ii) To show that Bd $C^n = S^{n-1}$, it suffices to show that π_i (Bd C^n) $= 0$ for $0 \leqslant i \leqslant n-1$ and then apply the generalized Poincaré conjecture. Let D be an $(i+1)$-cell whose boundary is the i-sphere T, $0 \leqslant i \leqslant n-1$ and let f be a continuous map of T into Bd C^n. Again by Brown's collaring theorem there exists a homomorphism h: Bd $C^n \times [0,1] \longrightarrow C^n$ such that $h(x,0) = x$ for each $x \in$ Bd C^n. Let A be a tame n-cell in int $C^n = E^n$ such that $h(\text{Bd } C^n \times \{1\}) \subset \text{int } A$. Choose $\varepsilon > 0$ so small that $h(\text{Bd } C^n \times \{t\}) \subset \text{int } C^n - A$, $0 \leqslant t \leqslant \varepsilon$. Now int $C^n - A \subset h(\mathrm{B} \times [0,1])$ and int C^n-A is homomorphic to $S^{n-1} \times E^1$. Hence $\pi_i(\text{int } C^n - A) = 0$ if $0 \leqslant i \leqslant n-1$. Let $g: T \longrightarrow \text{int } C^n - A$ be defined as $g(x) = h(f(x), \varepsilon)$. Now g can be extended to a continuous map of D into int C^n-A. Let r be the deformation retraction of $h(\text{Bd } C^n \times [0,1])$ onto Bd $C^n \subset C^n$ by $r(h(x,t)) = x \in$ Bd C^n. Then the map $r \circ g$: $D \longrightarrow$ Bd C^n is an extension of the map f taking T into Bd C^n. Therefore, Bd C^n is a combinatorial homotopy sphere of dimension $n-1 \geqslant 5$ and is therefore homomorphic to S^{n-1} by the generalized Poincaré conjecture.

PROPOSITION II.17. *If M^n is a compact contractible combinatorial n-manifold with boundary ($n \geqslant 4$), then $2M^n = S^n$ if and only if $M^n \times I = I^{n+1}$ (where $2M^n$ is two copies of M^n identified along their boundaries).*

Proof. It suffices to prove that $\text{Bd}\,(M^n \times I) = 2M^n$. For if $M^n \times I = I^{n+1}$, then $S^n = \text{Bd}\, I^{n+1} = \text{Bd}\,(M^n \times I) = 2M^n$. If $2M^n = S^n$, then $S^n = 2M^n = \text{Bd}\,(M^n \times I)$ and since $n+1 \geqslant 5$, it follows by Proposition II.16 that $M^n \times I = I^{n+1}$. To prove $\text{Bd}\,(M^n \times I) = 2M^n$ we use Brown's collaring theorem again. That is, there exists $\hat{h}: \text{Bd}\, M^n \times [0,1) \longrightarrow M^n$ so that $\hat{h}(m,0) = m \in \text{Bd}\, M^n$. Let i_1, i_2 be the inclusion maps taking M homomorphically onto each copy of M in $2M^n$. Let j_0, j_1 be the inclusion maps taking M homomorphically onto $M^n \times \{0\}$ and $M^n \times \{1\}$ respectively in $\text{Bd}\,(M^n \times I)$. Let $h: \text{Bd}\, M^n \times (-1, 1) \longrightarrow 2M^n$ be defined as

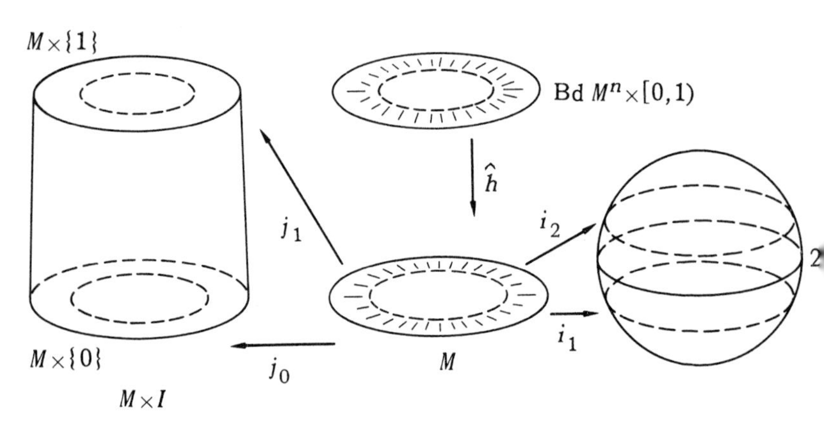

$$h(m, t) = \begin{cases} i_1 \circ \hat{h}(m, |t|) & -1 < t \leqslant 0 \\ i_2 \circ \hat{h}(m, t) & 0 \leqslant t < 1 \end{cases}.$$ Let $\hat{H}$ be the homomorphism taking $h(\text{Bd } M^n \times [-\frac{1}{2}, \frac{1}{2}]) \subset 2M^n$ into $\text{Bd }(M^n \times I) = M^n \times \{0\} \cup (\text{Bd } M^n) \times I \cup M^n \times \{1\}$ defined as follows:

$$\hat{H}(h(m, t)) = \begin{cases} j_0 \circ \hat{h}(m, -\frac{1}{2} - 2t) & -\frac{1}{2} \leqslant t \leqslant -\frac{1}{4} \\ m \times \{2t + \frac{1}{2}\} & -\frac{1}{4} \leqslant t \leqslant \frac{1}{4} \\ j_1 \circ \hat{h}(m, 2t - \frac{1}{2}) & \frac{1}{4} \leqslant t \leqslant \frac{1}{2} \end{cases}$$

Finally, define H taking $2M$ onto $\text{Bd}(M^n \times I)$ by

$$H(x) = \begin{cases} j_0 \circ i_1^{-1}(x) \text{ for } x \in i_1(M) - h(\text{Bd}(M^n \times [-\frac{1}{2}, \frac{1}{2}]) \\ \hat{H}(x) \text{ for } x \in h(\text{Bd } M^n \times [-\frac{1}{2}, \frac{1}{2}]) \\ j_0 \circ i_2^{-1}(x) \text{ for } x \in i_2(M) - h(\text{Bd } M^n \times [-\frac{1}{2}, \frac{1}{2}]). \end{cases}$$

THEOREM II.18. *If M^n is a contractible combinatorial n-manifold and $n \geqslant 5$, then $M^n \times I$ is homomorphic with I^{n+1}.*

Proof. Since M^n is contractible, $2M^n$ is a homotopy combinatorial n-sphere (i.e., by van Kampen's theorem, $\pi_1(2M^n) = 0$, by the Mayer-Vietoris sequence $H_i(2M) = 0$, $1 \leqslant i \leqslant n-1$, and by the Hurewicz theorem $\pi_i(2M) = 0$, $1 \leqslant i \leqslant n-1$). Since $n \geqslant 5$, it follows by the Poincaré conjecture that $2M^n = S^n$. Therefore, applying Proposition II.17, we have that $M^n \times I = I^{n+1}$.

THEOREM II.19. *If M^p and N^q are pseudo cells of positive dimensions p and q respectively, with $p + q \geqslant 6$, then $M^p \times N^q = I^{p+q}$.*

Proof. Since M^p and N^q are pseudo cells, int M^p and int N^q are contractible open combinatorial manifolds. By Stallings's result, int $(M^p \times N^q)$ = int $M^p \times$ int N^q is pwl homomorphic to E^{p+q} (here all is needed is that $p + q \geqslant 5$). By Proposition II.16, since $p + q \geqslant 6$, $M^p \times N^q = I^{p+q}$.

COROLLARY II.20. *If $n \geqslant 8$, then I^n is the product of two combinatorial manifolds with boundary, neither of which is a topological cell.*

This result follows immediately from Theorem II.19 after we have demonstrated the existence of pseudo n-cells, $n \geqslant 4$, which are not topological cells.

LEMMA II.21. *If P is a k-dimensional subcomplex contained in the interior of a combinatorial n-manifold M and N is a regular neighborhood of P and $P \subset \text{int } N$, then $N\text{-}P \approx \text{Bd } N \times [0,1)$.*

Proof. What we are actually going to prove is that there is a combinatorial map φ: $\text{Bd } N \longrightarrow P$ such that N is combinatorially equivalent to $I \times \text{Bd} N \cup_{\varphi} P$, the mapping cylinder of φ. Since any two regular neighborhoods of P in M are combinatorially equivalent by an equivalence carrying P onto P, it will suffice to prove the above statement for a particularly nice regular neighborhood of P, namely $N(|P''|, M'')$. First, let $N(|P'|, M')$ be the simplicial neighborhood of $|P'|$ under the first barycentric subdivision of M. For convenience let us denote this simplicial neighborhood by N'. Now there is a naturally defined retraction φ' of int N' onto P'. That is, each n-simplex $\sigma \in N'$, but not in P, can be expressed uniquely as the join of a simplex $\sigma_1 \in P'$ and a simplex $\sigma_2 \in \text{Bd} N'$. Hence, each point $x \in \text{int } \sigma$ lies on a unique segment $\langle p, n \rangle \subset \sigma$, where $p \in \sigma_1$ and $n \in \sigma_2$. The combinatorial retraction φ' of int N' onto P' is defined to take $x \in \langle p, n \rangle\text{-}\{n\}$ onto p and is the identity on P'.

Now we consider $N(|P''|, M'') \subset \text{int } N'$. The restriction $\varphi' \mid N(|P''|, M'') = \tilde{\varphi}'$ gives a combinatorial retraction of $N(|P''|, M'')$ onto P''. Let $\varphi = \tilde{\varphi}' \mid \text{Bd} N(|P''|, M'')$. Clearly, $N(|P''|, M'')$ is combinatorially equivalent to

$I \times \mathrm{Bd}\, N(|P''|, M'') \cup_{\varphi} P''$, the mapping cylinder of φ taking $\mathrm{Bd}\, N(|P''|, M'')$ onto P''.

THEOREM II.22. *Let P be a contractible k-dimensional subcomplex of a combinatorial n-sphere S^n and let N be a regular neighborhood of P having boundary B. Then*

(i) *if $n \leqslant 3$, the $N = I^n$,*

(ii) *if $n \neq 5$, and $k \leqslant n$-3 then $N = I^n$,*

(iii) *if $n \geqslant 4$, and n-2$\leqslant k \leqslant n$, then it may happen that $\pi_1(B) \neq 0$ and hence $N \neq I^n$.*

Proof of (i): If $n \leqslant 2$ then P must necessarily collapse to a point and hence N^n collapses to a point and therefore $N^n \approx I^n$. If $n = 3$, then $\mathrm{Bd}\, N^3$ is a homology 2-sphere and hence a real 2-sphere. Then by elementary results of 3-dimensional topology or by the generalised Schoenflies theorem it follows that $N^3 = I^3$.

Proof of (ii): Because of (i) we only need to consider $n \geqslant 6$. Since $k \leqslant n$-3 and N^n-$P \approx \mathrm{Bd}\, N^n \times [0,1)$, it follows that $\pi_1(\mathrm{Bd}\, N^n) = 0$. (One sees that every simple closed curve in N^n-P is trivial by considering any map of a disk into N having J as its boundary (since N is contractible we have this) and then apply the general position lemma. Since $k \leqslant n$-3, the general position image of the disk will miss P and hence J is trivial in N^n-P.) Also $\mathrm{Bd}\, N^n$ is a homology (n-1)-sphere (this follows from the Poincaré duality exact sequence for an n-manifold N with boundary: $\ldots \longrightarrow H^{q-1}(\mathrm{Bd}\, N) \longrightarrow H_{n-q}(N) \longrightarrow H^q(N) \longrightarrow H^q(\mathrm{Bd}\, N) \longrightarrow \ldots$). Hence since $\pi_1(\mathrm{Bd}\, N^n) = 0$ and $H_i(\mathrm{Bd}\, N^n) = 0$, $i \geqslant 1$, it follows that $\mathrm{Bd}\, N^n$ is in fact a homotopy (n-1)-sphere. Since n-1$\geqslant 5$, it follows by the Poincaré conjecture that $\mathrm{Bd}\, N = S^{n-1}$ and hence by Proposition II.16 of Volume I, we have that $N^n = I^n$.

Alternatively since $\pi_1(\text{Bd}\,N^n) = 0$ and N is contractible, we can use Stallings's result, i.e., int N is an open contractible n-manifold 1-connected at ∞. Hence int $N \approx E^n$ and by Proposition II.16, we again have $N^n = I^n$. (Note we have int $N \approx E^n$ even if $n = 5$.)

Remarks on (iii): This case will follow from results and examples to be discussed shortly. The examples of Poenaru [52] or Mazur [39] satisfy (iii) for $n=4$, $k=2$ and those of Curtis [17] for $k=n>4$. Each of these examples can be reduced by Whitehead elementary contractions to an example with $k=n-1$. Examples by Glaser [26] satisfy (iii) for $k=n-2$, $n-1$, and n for $n \geqslant 4$. The only unknown case is for contractible 2-complexes in S^5. In this case it can be easily shown that if one considers such an N^5, then int $N^5 \approx E^5$ and $\text{Bd}\,N^5$ is a homotopy 4-sphere, but it is still not known if such an N^5 is a 5-cell.

THEOREM II.23. *For $n \geqslant 5$, there exists countably many non-homomorphic pseudo n-cells N_i^n $(i=1,2,\dots)$ such that $N_i^n \neq I^n$ and $N_i^n \times I = I^{n+1}$ for all i.*

First, we prove two propositions from which Theorem II.23 will easily follow.

PROPOSITION II.24. *There exist countably many non-homomorphic homologically trivial 2-complexes each having distinct non-trivial fundamental groups.*

Proof. For each $n \geqslant 2$, let P_n be the presentation of the group G_n given by $P_n = \{a, b \mid a^{n-2} = (ab)^{n-1},\ b^3 = (ba^{-2}ba^2)^2\}$. Let K_n^2 be the 2-complex formed by attaching two disks D_1, D_2 to a figure-eight $a \vee b$ by the relations $a^{n-2}(ab)^{1-n}$, $b^3(ba^{-2}ba^2)^{-2}$ respectively. It then follows that $\pi_1(K_n^2) \cong G_n$ for all $n \geqslant 2$.

To see that $H_i(K_n^2) = 0$ for $i \geqslant 1$ and $n \geqslant 2$, we first observe that $H_1(K_n^2) = 0$ by adding the relation $ab = ba$ to each presentation and noting that the resulting group is trivial. To check $H_2(K_n^2)$, we just have to consider integral 2-chains of the form $C_2 = pD_1 + qD_2$, $p, q \in Z$. For C_2 to be a cycle we must have that $0 = \partial C_2 = p\partial D_1 + q\partial D_2 = p[(n\text{-}2)a + (1\text{-}n)a + (1\text{-}n)b] + q[3b\text{-}4b]$. That is $p(a+(n\text{-}1)b) + qb = 0$. But this follows only if $p = 0$ and $p = 0$ and hence K_n^2 has no 2-cycles.

We observe that for even $n \geqslant 6$, G_n is non-trivial by defining a homomorphism ψ taking G_n onto the alternating group A_n by $\psi(a) = (12)(34\ldots n)$ and $\psi(b) = (123)$. Later we will need the fact that G_n is indecomposable also. This follows since G_n is perfect as was shown above. That is, if $G_n = A*B$, the free product of two non-trivial groups A and B, then each of A and B are perfect since G_n is. Now $\{a, b\}$ is a minimal set of generators for G_n and hence a must be in one factor and b is in the other. But this leads to a perfect group on one generator which is impossible.

Finally, we claim that the sequence $G_6, G_8, G_{10}, \ldots$ contains infinitely many distinct groups. Let $\varphi: Z^+ \longrightarrow Z^+$ be the map of the natural numbers defined by $\varphi(n) = n! + 2$. We will show that $G_n, G_{\varphi(n)}, G_{\varphi^2(n)}, \ldots$ are all distinct by showing that there is no surjective map from $G_{\varphi^j(n)}$ to $G_{\varphi^i(n)}$ whenever $i < j$. First we consider $G_n, G_{\varphi(n)}$. *If* $\eta: G_{n!+2} \longrightarrow G_n$ is surjective, then so is $\psi \circ \eta: G_{n!+2} \longrightarrow A_n$. Now a and ab generate $G_{n!+2}$ so $\alpha = \psi \circ \eta(a)$ and $\beta = \psi \circ \eta(ab)$ generate A_n. But considering the presentation $P_{n!+2}$ we see that $\alpha^{n!} = \beta^{n!+1}$ and since the order of A_n is $n!/2$, β is the identity and we get the contradiction that the simple and non-abelian group A_n is generated by one element. Hence no surjective η exists.

Next suppose $\eta: G_{(n!+2)!+2} \longrightarrow G_n$ is surjective. This gives $\alpha^{(n!+2)!} = \beta^{(n!+2)!+1}$ in A_n and again $\beta = 1$. The case for $G_{\varphi^j(n)} \longrightarrow G_{\varphi^i(n)}$ goes in a similar way and thus it follows that $G_6, G_{\varphi(6)}, G_{\varphi^2(6)}, \ldots$ are all distinct.

PROPOSITION II.25. *If K is a homologically trivial 2-complex pwl embedded in S^n, $n \geqslant 5$ and if R is the regular neighborhood of K in S^n, then $\pi_1(K) = \pi_1(R) = \pi_1(\dot{R})$ and $\overline{S^n\text{-}R} = N^n$ is a pseudo n-cell such that $\pi_1(K) = \pi_1(\dot{N}^n)$.*

(Remark: this is false for $n = 4$ because in that case $\pi_1(\dot{R})$ depends on how K was embedded and entirely different techniques must be used in this dimension).

Proof. The fact that $\pi_1(K) = \pi_1(R)$ is trivial since K is a deformation retract of R. The fact that N^n is a combinatorial n-manifold with boundary is clear since R is itself (also clearly $\dot{N}^n = \dot{R}$). Since K has codimension $\geqslant 3$ in R and $R\text{-}K \approx \dot{R} \times [0,1)$, it follows that $\pi_1(K) = \pi_1(\dot{R}) = \pi_1(\dot{N}^n)$. Also since $n \geqslant 5$ it follows that $\pi_1(S^n\text{-}K) = 0$ and hence $\pi_1(N^n) = 0$ (N^n is a deformation of $S^n\text{-}K$). Since K is homologically trivial, it follows by Alexander duality that $H_i(S^n\text{-}K) = 0$ for $i \geqslant 1$ and hence $H_i(N^n) = 0$ for $i \geqslant 1$. Therefore $\pi_i(N^n) = 0$ for $i \geqslant 1$ and hence N^n is a pseudo n-cell.

Proof of Theorem II.23. For $n \geqslant 5$, the fact there exist countably many non-homomorphic pseudo n-cells N_i^n follows immediately from Propositions II.24 and II.25. The fact that $N_i^n \times I = I^{n+1}$ follows from Theorem II.18.

We now want to obtain a similar result corresponding to Theorem II.23 for $n = 4$. For the case $n = 4$ things are not as simple as we have already noted in the remark following Proposition

II.25. Also we do not have Theorem II.18 to fall back on to show that $N^n \times I = I^5$. The main results to be obtained are as follows:

THEOREM II.26. *There exist countably many different contractible 2-complexes P_i with regular neighborhoods $N_i^4 \subset S^4$ such that for every i:*

(1) $N_i^4 \times I \approx I^5$,

(2) $\pi_1(\mathrm{Bd}\, N_i^4) \neq 0$,

(3) $\pi_1(S^4 - P_i) \neq 0$;

and if $i \neq j$:

(4) $\pi_1(\mathrm{Bd}\, N_i^4) \neq \pi_1(\mathrm{Bd}\, N_i^4)$ *and hence* $N_i^4 \neq N_j^4$ *and* $\mathrm{int}\, N_i^4 \neq \mathrm{int}\, N_j^4$,

(5) $\pi_1(S^4 - P_i) \neq \pi_1(S^4 - P_j)$ *and hence* $S^4 - P_i \neq S^4 - P_j$.

THEOREM II.27. *For $n \geqslant 4$ there exist countably many different contractible $(n-2)$-complexes P_i^{n-2} with regular neighborhoods $M_i^n \subset S^n$ such that for every i:*

(1) $M_i^n \times I \approx I^{n+1}$,

(2) $\pi_1(\mathrm{Bd}\, M_i^n) \neq 0$,

(3) $\pi_1(S^n - P_i^{n-2}) \neq 0$;

and if $i \neq j$:

(4) $\pi_1(S^n - P_i^{n-2}) \neq \pi_1(S^n - P_j^{n-2})$.

Remark. Theorem II.27 gives the examples promised illustrating Property (iii) of Theorem II.22. We also note that in each of Theorems II.26 and II.27 we have that the regular neighborhoods crossed with an interval are combinatorial $n+1$ cells and that all the contractible $(n-2)$-complexes are pwl embedded in S^n so as to have non-simply connected complements.

Also each of the contractible $(n-2)$-complexes constructed in Theorems II.26 and II.27 can be embedded pwl in S^n so as to have simply connected complements and regular neighborhoods which are combinatorial n-cells.

THEOREM II.28. *For $n \geqslant 5$, the* M_i^n (of Theorem II.27) *are contractible combinatorial n-manifolds with boundary that are not topological n-cells, but are combinatorially equivalent to the union of two combinatorial n-balls which intersect in a combinatorial n-ball. Furthermore, for $n \geqslant 5$,* int $M_i^n \approx X \cup Y$ *where* $X \approx Y \approx X \cap Y \approx E^n$, *while* int $M_i^n \not\approx E^n$.

Theorem II.26 will be proven by first constructing countably many non-homomorphic 2-complexes and then constructing combinatorial 4-manifolds with boundary so as to be regular neighborhoods of the given 2-complexes. Theorem II.27 will be obtained by taking suspensions of the examples of Theorem II.26, and Theorem II.28 will follow by examining these suspensions more carefully. Now we give some details.

For each positive integer n, let K_n be the contractible 2-complex formed by attaching a disk D to a circle a by the formula $a^{n+1}a^{-n}$. We now want to embed K_n in a contractible 4-manifold W_n^4 with boundary so that W_n^4 can be considered as a regular neighborhood of K_n. Let T^3 be a solid 3-dimensional torus forming half of the boundary of $S^1 \times I^3$ in E^4. For each n, we will consider a certain embedding of a simple closed curve Γ_n in int T^3 so that:

(1) Γ_n is homotopic but not isotopic to the core $S^1 \times \{0\} \subset S^1 \times I^2 = T^3$;

(2) under the natural map $\rho\colon S^1 \times I^2 \longrightarrow S^1 \times \{0\}$, $\rho(\Gamma_n)$ wraps around $S^1 \times \{0\}$ $n+1$ times in a counterclockwise direction and n times in a clockwise direction;

(3) in forming the 4-manifold W_n^4 with boundary by attaching a 2-handle to $S^1 \times I^3$ along the curve Γ_n, we will get $\pi_1(\mathrm{Bd}\, W_n^4) \neq 0$.

More precisely the embeddings of Γ_1 and Γ_2 are illustrated below. The embedding of Γ_n, in general, is indicated in the following diagram.

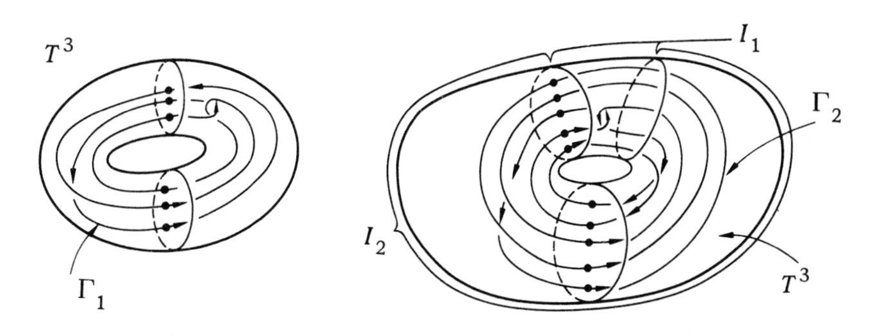

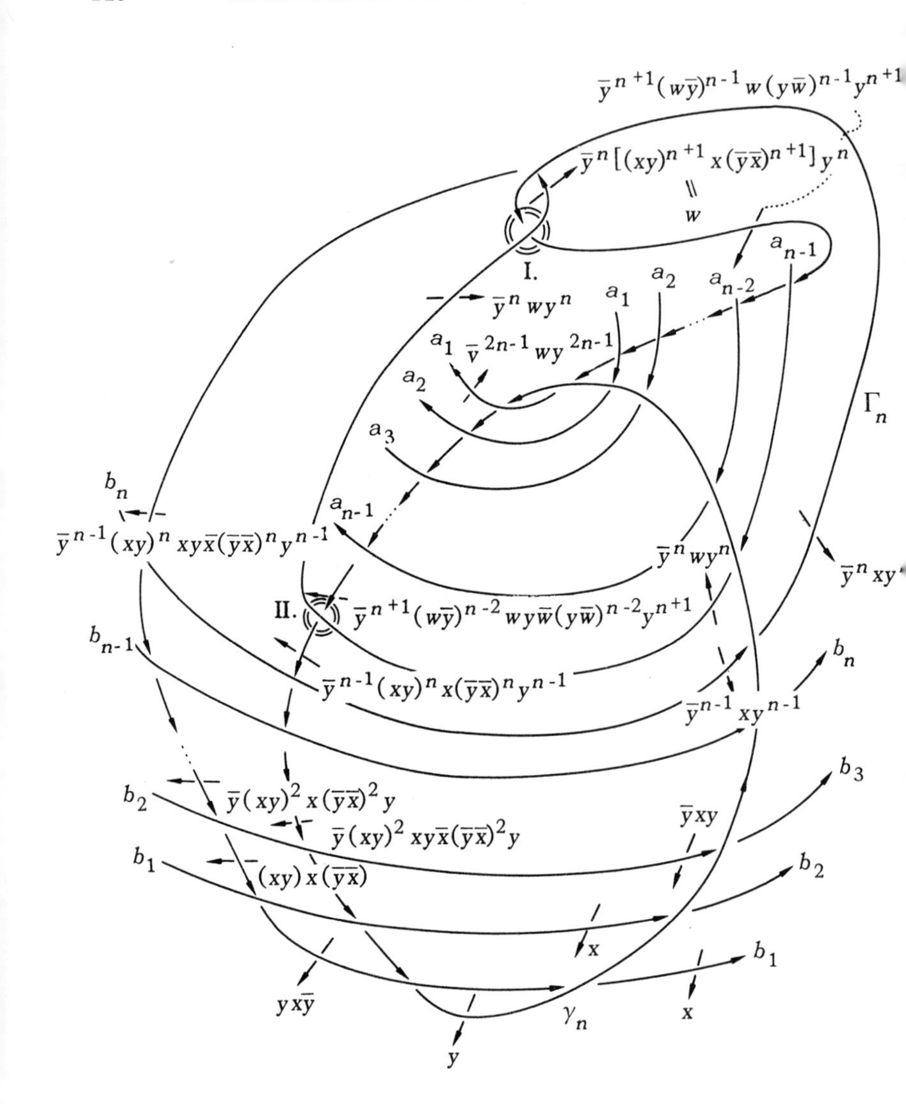

$\bar{y}^{n+1}(w\bar{y})^{n-1}w(y\bar{w})^{n-1}y^{n+1}$
$\bar{y}^{n}[(xy)^{n+1}x(\bar{y}\bar{x})^{n+1}]y^{n}$
$\|$
w
I.
a_{n-1}
a_1
a_2
a_{n-2}
$\bar{y}^{n}wy^{n}$
a_1
$\bar{v}^{2n-1}wy^{2n-1}$
a_2
a_3
Γ_n
b_n
a_{n-1}
$\bar{y}^{n-1}(xy)^{n}xy\bar{x}(\bar{y}\bar{x})^{n}y^{n-1}$
$\bar{y}^{n}wy^{n}$
$\bar{y}^{n}xy$
II.
$\bar{y}^{n+1}(w\bar{y})^{n-2}wy\bar{w}(y\bar{w})^{n-2}y^{n+1}$
b_{n-1}
b_n
$\bar{y}^{n-1}(xy)^{n}x(\bar{y}\bar{x})^{n}y^{n-1}$
$\bar{y}^{n-1}xy^{n-1}$
b_3
b_2
$\bar{y}(xy)^{2}x(\bar{y}\bar{x})^{2}y$
$\bar{y}xy$
$\bar{y}(xy)^{2}xy\bar{x}(\bar{y}\bar{x})^{2}y$
b_1
$(xy)x(\bar{y}\bar{x})$
b_2
x
b_1
$yx\bar{y}$
γ_n
x
y

For fixed n, consider $\Gamma_n \subset \text{int } T^3 \subset S^1 \times S^2 = \text{Bd}(S^1 \times I^3)$, where Γ_n is piecewise linearly embedded in int T^3 as indicated in the previous figures. Let N be a small regular neighborhood of Γ_n in $S^1 \times S^2$ so that $N \subset \text{int } T^3$. Now $N \approx S^1 \times I^2$. Let j be a pwl embedding of $\text{Bd}I^2 \times I^2$ into int T^3 such that $j(\text{Bd}I^2 \times \{0\}) = \Gamma_n$ ($0 \in \text{int } I^2$) and $j(\text{Bd}I^2 \times I^2) = N$. We define out 4-manifold W_n^4 with boundary as $I^2 \times I^2 \cup_j S^1 \times I^3$. Also for simplicity in computing $\pi_1(\text{Bd } W_n^4)$ we will assume that in adding this 2-handle, there is no 'twisting' of the tubular neighborhood thus formed as we go around Γ_n.

LEMMA II.29. *W_n^4 can be given a combinatorial triangulation so that K_n can be considered as a subcomplex of W^4 and $W_n^4 \setminus K_n$.*

Proof. First we consider the part of W_n^4 corresponding to $S^1 \times I^3$. Divide S^1 into two arcs I_1, I_2 such that $I_1 \times S^2$ contains the part of Γ_n where the curve is locally linked. Let $f: S^1 \longrightarrow S^1$ be a pwl map shrinking I_1 to a point. Let $p: S^1 \times S^2 \longrightarrow S^1$ be the projection onto the first factor, and let g, h be the compositions:

$$\Gamma_n \xrightarrow{\subset} S^1 \times S^2 \xrightarrow{p} S^1 \xrightarrow{f} S^1, \qquad g = f \circ p: S^1 \times S^2 \longrightarrow S^1, \qquad h: \Gamma_n \longrightarrow S^1$$

Let $M(g)$, $M(h)$ denote the mapping cylinders of g, h. We can extend the identity on $S^1 \times S^2$ to a pwl homomorphism $M(g) \approx S^1 \times I^3$ as follows: Let $g_1: I_1 \times S^2 \longrightarrow fI_1$ and $g_2; I_2 \times S^2 \longrightarrow S^1$ denote the restrictions of g. Regard the 4-ball $I_1 \times I^3$ as a cone on its boundary, with vertex v say. Map $M(g_1)$ homomorphically onto the subcone $v(I_1 \times S^2)$, and extend to a homomorphism of $M(g_2)$ onto $v(\dot{I}_1 \times I^3) \cup I_2 \times I^3$. Therefore we can write $M(g) = S^1 \times I^3$.

Now $M(h)$ is a subcylinder of $M(g)$ since $h = g|\Gamma_n$, hence $M(g)\searrow M(h)$ because a mapping cylinder collapses onto any subcylinder. The 2-handle $I^2 \times I^2$ of W_n^4 collapses to $I^2 \times \{0\}$, which is a disk spanning Γ_n. Therefore,

$$W_n^4 \searrow I^2 \times \{0\}_j \cup \underbrace{S^1 \times I^3 \searrow I^2 \times \{0\}_j \cup M(h)}_{M(g)}.$$

But $I^2 \times \{0\}_j \cup M(h)$ is piecewise linearly homomorphic to K_n.

Lemma II.30. *Any two pwl embeddings of a contractible 2-complex K in E^5 are combinatorially equivalent and hence their regular neighborhoods are pwl homomorphic.*

Remark. In fact the following is true (and is proved in a similar manner). Any two pwl embeddings of a k-complex K in E^{2k+1} are equivalent under a pwl isotopy if $k \geqslant 2$ and $H^k(K) = 0$. Also for all integers k and n with $k \geqslant 2$ and $k \leqslant n \leqslant 2k$, there is a collapsible k-complex $K(k, n)$ such that $K(k,n)$ can be embedded in two non-equivalent ways in E^n. Furthermore, it can be shown that any two pwl embeddings of any 2-complex in a combinatorial 5-manifold have pwl homomorphic regular neighborhoods.

First, we note that the reason we want the above result is because of the following two corollaries.

Corollary II.31. *Suppose K is a contractible 2-subcomplex in the interior of a combinatorial 4-manifold M and N is a regular neighborhood of K in M. If K can be combinatorially embedded in E^3, then N can be embedded in E^4 and $N \times I \approx I^5$.*

Proof. By Theorem II.18, $N \times I^2 = I^6$. Since $\text{Bd}\,(N \times I^2)$ is homomorphic to S^5 triangulated as a combinatorial 5-manifold and $2(N \times I) \approx \text{Bd}(N \times I^2)$, $N \times I$ can be combinatorially embedded in a combinatorial triangulation of E^5. Let $\hat{K}$ be a combinatorial embedding of K in $E^3 \subset E^5$. Since the regular neighborhood of $\hat{K}$ in E^3 is necessarily a combinatorial 3-ball, the regular neighborhood $\hat{N}$ of $\hat{K}$ in E^5 is then a combinatorial 5-ball. Hence by Lemma II.30, $\hat{N} \approx N \times I \approx I^5$. The fact that $2N \approx \mathrm{Bd}(N \times I) \approx S^5$ gives that N can be combinatorial embedded in E^4.

COROLLARY II.32. *The* W_n^4 *of* Lemma II.29 *can be combinatorially embedded in* S^4 *and* $W_n^4 \times I \approx I^5$.

Proof. We observe that for each n, K_n can be embedded in S^3, and hence the result follows by Corollary II.31.

Proof of Lemma II.30: Suppose K is a 2-complex and K_1, K_2 are two pwl embeddings of K in a combinatorial triangulation of E^5. That is, we have f_1, f_2 pwl homomorphisms carrying K into E^5, where $f_1(K) = K_1$ and $f_2(K) = K_2$. Let $K^{(1)}$ be the 1-skeleton of K. Then $f_2 \circ f_1^{-1}|_{f_1(K^{(1)})}$ is a pwl homomorphism carrying $K_1^{(1)}$ onto $K_2^{(1)}$. By the results of Section D, Chapter IV in Volume I, there exists a a pwl homomorphism h_1 of E^5 onto itself such that $h_1|_{f_1(K^{(1)})} = f_2 \circ f_1^{-1}$. Let us subdivide $h_1(K_1)$ so as to be a simplicial complex. Let h_2 be a homomorphism of E^5 onto itself leaving $h_1(K_1^{(1)})$ fixed so that the other vertices of $h_1(K_1)$ are in general position and also in general position with respect to K_2. Then $h_2(h_1(K_2)) \cup K_2$ *is a* 2-complex in E^5 such that $h_2(h_1(K_1)) \cap K_2 = K_2^{(1)}$ and for each pair of 2-dimensional simplexes $\sigma, \sigma' \in K$ (not necessarily distinct) $h_2(h_1(f_1(\mathrm{int}\,\sigma))) \cap f_2(\mathrm{int}\,\sigma') = \emptyset$. Also for each 2-simplex $\sigma \in K$, we have that $h_2(h_1(f_1(\sigma))) \cup f_2(\sigma)$ is a pwl 2-sphere in E^5, and for $x \in \sigma$, $h_2(h_1(f_1(x))) = f_2(x)$. Let $\sigma_1, \sigma_2, \ldots, \sigma_r$ be the 2-simplexes of K.

Now we have the result that any pwl 2-sphere in E^5 is equivalent to the boundary of a 3-simplex in E^5. That is, we can find a pwl homomorphism h'_{σ_1} taking E^5 onto E^5 such that h'_{σ_1} $[h_2(h_1(f_1(\sigma_1))) \cup f_2(\sigma_1)] = \dot{\Delta}^3 \subset E^5$ and so that h'_{σ_1} $[(h_2 \circ h_1 \circ f_1(K)) \cup f_2(K)]$ is in general position with respect to Δ^3. Let $h'_{\sigma_1}(h_2 \circ h_1 \circ f_1(\sigma_1)) = A^2$ and $h'_{\sigma_1}(f_2(\sigma_1)) = B^2$. There exists a 3-simplex $\tilde{\Delta}^3 \subset \text{int}\, \Delta^3$ such that $(\text{int}\, \Delta^3 \text{-int}\, \tilde{\Delta}^3) \cap h_{\sigma_1}$ $(h_2 \circ h_1(K_1) \cup K_2) = \emptyset$. Now $E^5 - h'_{\sigma_1}(K_2) \approx E^5 - \{\text{point}\}$, since K_2 is contractible. Hence $\mathring{\tilde{\Delta}}^3$ bounds a polyhedral 3-ball D^3 in $E^5 - h'_{\sigma_1}(K_2)$. Also $\Delta^3\text{-int}\, \tilde{\Delta}^3 \approx \dot{\Delta}^3 \times [0,1] = A^2 \times [0,1] \cup B^2 \times [0,1]$. $\mathring{D}^3 = (A^2 \times \{1\}) \cup (B^2 \times \{1\}) = \tilde{A}^2 \cup \tilde{B}^2$. Let the annulus $\mathring{A}^2 \times [0,\dot{1}] = C$. There is a pwl homomorphism g_1 taking E^5 onto E^5, fixed on $h'_{\sigma_1}(K_2)$, taking A^2 onto $C \cup \tilde{A}^2$ using the 3-ball $A^2 \times [0,1]$. There is a $g_2: E^5 \longrightarrow E^5$, fixed on $h'_{\sigma_1}(K_2)$, taking $g_1(A^2)$ onto $C \cup \tilde{B}^2$ using the 3-ball D^3. Finally there is a $g_3: E^5 \longrightarrow E^5$, fixed on $h'_{\sigma_1}(K_2 - f_2(\mathring{\sigma}_1))$ taking $g_2 \circ g_1(A^2)$ onto B^2. Let $h_{\sigma_1} =$ $h'^{-1}_{\sigma_1} \circ g_3 \circ g_2 \circ g_1 \circ h'_{\sigma_1}$. Then h_{σ_1} is a pwl homomorphism carrying E^5 onto itself, fixed on $K_2 - f_2(\mathring{\sigma}_1)$ and carrying $h_2 \circ h_1 \circ f_1(\mathring{\sigma}_1)$ onto $f_2(\sigma_1)$. We now apply the same technique to $h_{\sigma_1} \circ h_2 \circ h_1 \circ f_1(\sigma_2)$ $\cup f_2(\sigma_2)$ to get $h_{\sigma_2}: E^5 \longrightarrow E^5$, fixed on $K_2 - f_2(\mathring{\sigma}_2)$ and carrying $h_{\sigma_1} \circ h_2 \circ h_1 \circ f_1(\sigma_2)$ onto $f_2(\sigma_2)$. Inductively then, we can find a pwl homomorphism $h_{\sigma_1}: E^5 \longrightarrow E^5$, fixed on $K_2 - f_2(\sigma_1)$ taking $h_{\sigma_{1-1}} \circ \cdots \circ h_{\sigma_1} \circ h_2 \circ h_1 \circ f_1(\sigma_i)$ onto $f_2(\sigma_i)$. Thus $H =$ $h_{\sigma_r} \circ \cdots \circ h_{\sigma_1} \circ h_2 \circ h_1$ is a pwl homomorphism of E^5 onto itself carrying K_1 onto K_2.

We now go back to the regular neighborhoods W^4_n of K_n (by Lemma II.29) we constructed earlier. We now have to apply some techniques of knot theory (refer to Appendix B).

LEMMA II.33. *For each even n,* $\pi_1(\text{Bd}\, W_n^4) \neq 0$.

Proof. To find a presentation of $\pi_1(\text{Bd}\, W_n^4)$ we need to consider how $\text{Bd}\, W_n^4$ can be obtained from the 3-sphere S^3. The claim is that a presentation of $\pi_1(\text{Bd}\, W_n^4)$ can be obtained by looking at the fundamental group of $E^3\text{-}(\Gamma_n + \gamma_n)$ as indicated in the previous figure and adding two appropriate relations. Let $T(\gamma_n)$ and $T(\Gamma_n)$ denote disjoint 3-dimensional solid tori in S^3 having γ_n and Γ_n as cores respectively. We can think of $T(\gamma_n)$ and $T(\Gamma_n)$ as small tubular neighborhoods of γ_n and Γ_n. Let $M(\gamma_n)$, $L(\gamma_n)$, $M(\Gamma_n)$ and $L(\Gamma_n)$ denote nice meridional and longitudinal curves in each of $\text{Bd}\, T(\gamma_n)$ and $\text{Bd}\, T(\Gamma_n)$ respectively. We can think of each of $L(\gamma_n)$ and $L(\Gamma_n)$ as the curves γ_n and Γ_n displaced slightly so that neither of these curves twist around their respective cores.

We recall that W_n^4 was obtained by adding a 2-handle to $S^1 \times I^3$. That is, $W_n^4 = I^2 \times I^2 \cup_j S^1 \times I^3$ where $j(\text{Bd}\, I^2 \times I^2) = N$ and N was a small tubular neighborhood of Γ_n in $\text{Bd}(S^1 \times I^3)$. Thus $\text{Bd}\, W_n^4 =$ $[\text{Bd}(I^2 \times I^2)\text{-int}\,(\text{Bd}\, I^2 \times I^2)] \cup_{j'} [\text{Bd}(S^1 \times I^3)\text{-int}\, N]$ where $j' = j|_{\text{Bd}\, I^2 \times \text{Bd}\, I^2}$. Since the first half of the above formula is just $I^2 \times \text{Bd}\, I^2$, $\text{Bd}\, W_n^4$ is obtained by removing the interior of a 3-dimensional torus in $S^1 \times S^2$ having Γ_n as a core, namely N, and then sewing back in another 3-dimensional torus $I^2 \times \text{Bd}\, I^2$. In sewing back in $I^2 \times \text{Bd}\, I^2$ we have identified a nice meridional curve in $\text{Bd}(I^2 \times \text{Bd}\, I^2)$ with a nice longitudinal curve in $\text{Bd}\, N$ and a nice longitudinal curve in $\text{Bd}(I^2 \times \text{Bd}\, I^2)$ with a nice meridional curve in $\text{Bd}\, N$. That is in forming $\text{Bd}\, W_n^4$ we essentially removed a solid 3-dimensional torus from $S^1 \times S^2 = \text{Bd}(S^1 \times I^3)$ and then 'sew' another one in 'backwards without any twisting'.

Now let us return to $T(\gamma_n)$ and $T(\Gamma_n)$ lying in S^3. If we remove $T(\gamma_n)$ from S^3 and then sew it in backwards, we obtain a

3-manifold homomorphic to $S^1 \times S^2$. $T(\Gamma_n)$ is then embedded in this copy of $S^1 \times S^2$ in exactly the same fashion as N is embedded in $\mathrm{Bd}(S^1 \times I^2)$ used in forming W_n^4. Now by removing $T(\Gamma_n)$ and then sewing it in backwards we essentially obtain $\mathrm{Bd}\, W_n^4$. Hence to find a presentation of $\pi_1(\mathrm{Bd}\, W_n^4)$ we merely have to obtain a presentation for $\pi_1(E^3 - (T(\gamma_n) + T(\Gamma_n))$ and then apply Van Kampen's theorem twice, once after sewing $T(\gamma_n)$ in backwards and then after sewing $T(\Gamma_n)$ in backwards. Sewing $T(\gamma_n)$ in backwards corresponds to trivializing the curve corresponding to $L(\gamma_n)$ in our presentation of $\pi_1(E^3 - (T(\gamma_n) + T(\Gamma_n))$. Sewing $T(\Gamma_n)$ in backwards corresponds to trivializing the curve corresponding to $L(\Gamma_n)$.

Thus a presentation of $\pi_1(\mathrm{Bd}\, W_n^4)$ can be obtained by looking at the fundamental group of $E^3 - (\gamma_n + \Gamma_n)$ (since $\pi_1(E^3 - (\gamma_n + \Gamma_n)) \cong \pi_1(E^3 - (T(\gamma_n) + T(\Gamma_n)))$ as in the previous figure and adding the relations corresponding to curves slightly above each of γ_n and Γ_n respectively. Adding the relation above γ_n corresponds to trivializing $L(\gamma_n)$ and gives us a presentation for $\pi_1(S^1 \times S^2 - \Gamma_n)$. Adding in the relation above Γ_n corresponds to trivializing $L(\Gamma_n)$ and brings the 2-handle of W_n^4 into consideration. We then obtain a presentation for $\pi_1(\mathrm{Bd}\, W_n^4)$. We note the generators y and x indicated in the figure correspond to $M(\gamma_n)$ and $M(\Gamma_n)$ in a presentation for $\pi_1(E^3 - (T(\gamma_n) + T(\Gamma_n))$.

The resulting group has the following presentation:

generators: x and y

Relations: $w = (xy)^{n+1} x (\overline{yx})^{n+1}$ (where $\bar{x}$ denotes x^{-1})

and

I: $\bar{x}(w\bar{y})^n w (y\bar{w})^n = 1,$

II. $(xy)^{n+1} y (\bar{y}\bar{x})^{n+1} \bar{y} (w\bar{y})^n \bar{w} (y\bar{w})^n w = 1,$

γ_n: $\bar{y}^{2n+1} (y\bar{w})^n y (xy)^n x = 1,$

Γ_n: $(w\bar{y})^n (xy)^{n+1} = 1.$

Using I. and $w = (xy)^{n+1}x(\overline{y}\overline{x})^{n+1}$ in II. we get that $1 = 1$. Using the fact that Γ_n : gives $(xy)^{n+1} = (y\overline{w})^n$, relation I. is equivalent to $w = (xy)^{n+1}x(\overline{y}\overline{x})^{n+1}$. Hence we get that $\pi_1(\mathrm{Bd}\,W_n^4)$ has the following presentation:

$$G_n = \{x, y \mid y^{2n+2} = (xy)^{n+1}y(xy)^{n+1},\ (xy)^{n+1} = [y\underbrace{(xy)^{n+1}\overline{x}(\overline{y}\overline{x})^{n+1}}_{\overset{\|}{w}}]^n\}.$$

Adding the relation $(xy)^{n+1} = 1$, we get:

$$G_n' = \{x, y \mid (xy)^{n+1} = y^{2n+1} = (y\overline{x})^n = 1\}.$$

Setting $xy = \beta$ and $y^2 = \alpha$, we get the group:

$$G_n'' = \{\alpha, \beta \mid \beta^{n+1} = \alpha^{2n+1} = (\alpha\overline{\beta})^n = 1\}.$$

If we consider the groups for n even, we can add the relation $(\alpha\overline{\beta})^2 = 1$ and hence we have:

$$\hat{G}_n = \{\alpha, \beta \mid \beta^{n+1} = \alpha^{2n+1} = (\alpha\overline{\beta})^2 = 1\}.$$

$\hat{G}_n$ can be shown to have a non-trivial representation in the alternating group A_{2n+1} by sending $\alpha \to (123, \ldots, 2n, 2n+1)$ and $\beta \to (125, \ldots, 2n-1, 2n+1)$.

For each even n we have a map ψ_n' taking G_n onto $\hat{G}_n$ and a map ψ_n'' taking $\hat{G}_n$ into A_{2n+1}. Let $\hat{A}_{2n+1} = \psi_n''(\hat{G}_n)$ be the subgroup of A_{2n+1} generated by $(123, \ldots, 2n, 2n+1)$ and $(135, \ldots, 2n-1, 2n+1)$. Let $\psi_n = \psi_n'' \circ \psi_n'$ be the map taking G_n onto $\hat{A}_{2n+1}$. We note that $\hat{A}_{2n+1}$ is not abelian.

LEMMA II.34. *The sequence $G_2, G_4, G_6, \ldots$ contains infinitely many distinct groups.*

Proof. Let φ: $Z^+ \longrightarrow Z^+$ be the map of the positive integers defined by $\varphi(n) = (2n+1)!$. The proof is now similar to that given for Proposition II. 24. That is, we will show that G_n, $G_{\varphi(n)}$, $G_{\varphi^2(n)}$, ... are all distinct by showing that there is no surjective map.

$$G_{\varphi^j(n)} \longrightarrow G_{\varphi^i(n)}$$

whenever $i < j$.

In using the proof given earlier, we first look at G_n, $G_{\varphi(n)}$ as was done there. We note that if η: $G_{(2n+1)!} \longrightarrow G_n$ is surjective, then $\psi_n \circ \eta$: $G_{(2n+1)!} \longrightarrow \hat{A}_{2n+1}$ is surjective. Also we use the fact that y and xy generate $G_{(2n+1)!}$ and hence $v = \psi_n\, \eta\,(y)$ and $u = \psi_n\, \eta\,(xy)$ generate $\hat{A}_{2n+1}$. But in considering the relations defining $G_{(2n+1)!}$ we get that

$$u^{(2n+1)!+1} = [vu^{(2n+1)!+1}\, v\bar{u}^{(2n+1)!+2}]^{(2n+1)!}.$$

Since the order of A_{2n+1} containing $\hat{A}_{2n+1}$ is $^{(2n+1)!}/2$ we get that $u=1$ and hence we get the contradiction that the non-abelian group $\hat{A}_{2n+1}$ is generated by one element. The remainder of the proof is similar to that given earlier, except using the notation given above.

We now can obtain the proof of Theorem II.26: applying Lemma II.29, Corollary II.32, Lemma II.33 and Lemma II.34 we obtain countably many different contractible 2-complexes $K_i^2 \subset W_i^4 \subset E^4$ such that $W_i^4 \times I \approx I^5$ for every i and if $i \neq j$, $\pi_1(\mathrm{Bd}\, W_i^4) \neq \pi_1(\mathrm{Bd}\, W_j^4)$ (and both are not zero). Let P_i be the contractible 2-complex formed as the union of two copies of K_i^2 in $2W_i^4 \approx S^4$ plus a polygonal arc A_i intersecting each copy of K_i^2 and $\mathrm{Bd}\, W_i^4$ in a single point, respectively. Then we have that $\pi_1\,(S^4\text{-}P_i) = \pi_1(\mathrm{Bd}\, W_i^4) \neq 1$ since $\pi_1\,(S^4\text{-}P_i) \cong \pi_1\,(S^4\text{-}(K_{i_1} + K_{i_2}))$ and $S^4\text{-}(K_{i_1} + K_{i_2}) \approx (W_i^4 - K_{i_1}) \cup (W_{i_2}^4 - K_{i_2}) \approx \mathrm{Bd}\, W_i^4 \times (-1, 1)$. Now the regular neighborhood N_i^4 of P_i under the second barycentric subdivision of S^4 is $N(|K''_{i_1}|, S^{4''}) \cup N(|A''_i|, S^{4''}) \cup N(|K''_{i_2}|, S^{4''})$ and is combinatorially equivalent to two copies of W_i^4 identified together along a 3-ball in the boundary of each and hence $\pi_1(\mathrm{Bd}\, N_i^4) \cong \pi_1(\mathrm{Bd}\, W_i^4) * \pi_1(\mathrm{Bd}\, W_i^4)$. Also the fact that $N_i^4 \times I \approx I^5$ follows since $W_i^4 \times I \approx I^5$ and $N_i^4 \times I$ is combinatorially equivalent to two copies of $W_i^4 \times I$ identified along a 4-ball in the boundary of each. Hence Theorem II.26 follows.

Proof of Theorems II.27 and II.28. Let $\{N_i^4\}$ be the combinatorial 4-manifolds with boundary obtained in Theorem II.26. For each i we will consider $N_i^4 \subset S^4$ as in the above proof of Theorem II.26. Also for each i we have that $N_i^4 \searrow P_i^2$, $N_i^4 \times I \approx I^5$, $\pi_1(\mathrm{Bd}\, N_i^4) \neq 0$ and $\pi_1(S^4\text{-}P_i^2) \neq 0$ (also for $i \neq j$, $\pi_1(S^4\text{-}P_i)$ $\neq \pi_1(S^4\text{-}P_j)$). These examples satisfy Theorem II.27 for $n=4$ and will be used to construct examples showing that Theorems II.27 and II.28 are true for $n=5$. Then the new examples will allow us, by repeating exactly the same construction given below, to obtain examples for $n=6$ and hence by induction we get the desired results for $n \geqslant 5$. Hence for fixed i, let us consider the following construction:

(1) Let B^4 be a combinatorial 4-ball in S^4 such that $N_i^4 \subset \mathrm{int}\, B^4$. Then we have $\pi_1(S^4\text{-}P_i^2) = \pi_1(B^4\text{-}P_i^2) \neq 0$.

(2) Let ΣP_i^2 be the contractible 3-complex obtained from the suspension of P_i^2 (say from points p and q, i.e., $\Sigma P_i^2 \cong p P_i^2 q$) embedded in S^5 where $S^5 \cong \Sigma B^4 \cup \mathcal{C}(\mathrm{Bd}\, \Sigma B^4)$, with $\mathcal{C}(\mathrm{Bd}\, \Sigma B^4)$ the cone over the boundary of the 5-ball ΣB^4 – say from the point r. Thus $S^5 \cong p B^4 q + r(p \mathring{B}^4 q)$. Using this embedding of ΣP_i^2 (which will be our P_i^3) we have that $S^5\text{-}(P_i^3 + (p r q))$ is of the same homotopy type as $B^4\text{-}P_i$. Hence $\pi_1(S^5\text{-}P_i^3) = \pi_1(S^5\text{-}(P_i^3 + (p r q)))$ $= \pi_1(B^4\text{-}P_i) = \pi_1(S^4\text{-}P_i)$. Thus we have parts (3) and (4) of Theorem II.27 for $n=5$.

(3) Let M_i^5 be the regular neighborhood of P_i in S^5. Since $\pi_1(S^5\text{-}P_i) \neq 0$, it follows that $\pi_1(\mathrm{Bd}M_i^5) \neq 0$ (i.e. using van Kampen's theorem on $S^5 = M_i^5 \cup S^5\text{-int } M_i^5$ with $M_i^5 \cap (S^5\text{-int } M_i^5)$ $= \mathrm{Bd}M_i^5$; since M_i^5 is contractible, if $\pi_1(\mathrm{Bd}M_i^5) = 0$, this would imply that $\pi_1(S^n) \cong \pi_1(M_i^5) * \pi_1(S^5\text{-int}M_i^5) \cong \pi_1(S^5\text{-int}M_i^5)$ $\cong \pi_1((S^5\text{-int}M_i^5) \cup \mathrm{Bd}M_i^5 \times [0,1)) \cong \pi_1(S^5\text{-}P_i) \neq 0$, hence leading to a contradiction!). Thus M_i^5 is a pseudo 5-cell $\neq I^5$ and we have part (2) for $n=5$.

(4) $M_i^5 \approx B_1^5 \cup B_2^5$ where $B_1^5 \cap B_2^5 \approx I^5$ (this will give the first statement of Theorem II.28 for $n=5$). First we note that $M_i^5 \approx N(|P_i^{3\,\prime\prime}|, S^{5\,\prime\prime}) \approx N(|(pP_i^2)''|, S^{5\,\prime\prime}) \cup N(|(P_i^2 q)''|, S^{5\,\prime\prime}) \approx$ $B_1^5 \cup B_2^5$ since the cones pP_i^2 and $P_i^2 q$ collapse to a point. Also $N(|(pP_i^2)''|, S^{5\,\prime\prime}) \cap N(|(P_i^2 q)''|, S^{5\,\prime\prime}) = N(|P_i^{2\,\prime\prime}|, S^{5\,\prime\prime}) \approx N_i^4 \times I \approx I^5$. Thus $B_1^5 \cap B_2^5 \approx I^5$. By taking int $B_1^5 = X$ and int $B_2^5 = Y$ we get the second statement of Theorem II.28 for $n=5$. Int $M_i^5 \neq E^5$, since $\pi_1(\mathrm{Bd}M_i^5) \neq 0$ and $\mathrm{Bd}M_i^5$ is collared in M_i^5. Hence there will be nontrivial curves in a collar that cannot shrink to a point in the collar. But if int $M_i^5 = E^5$, curves 'near infinity' could be shrunk 'near infinity'.

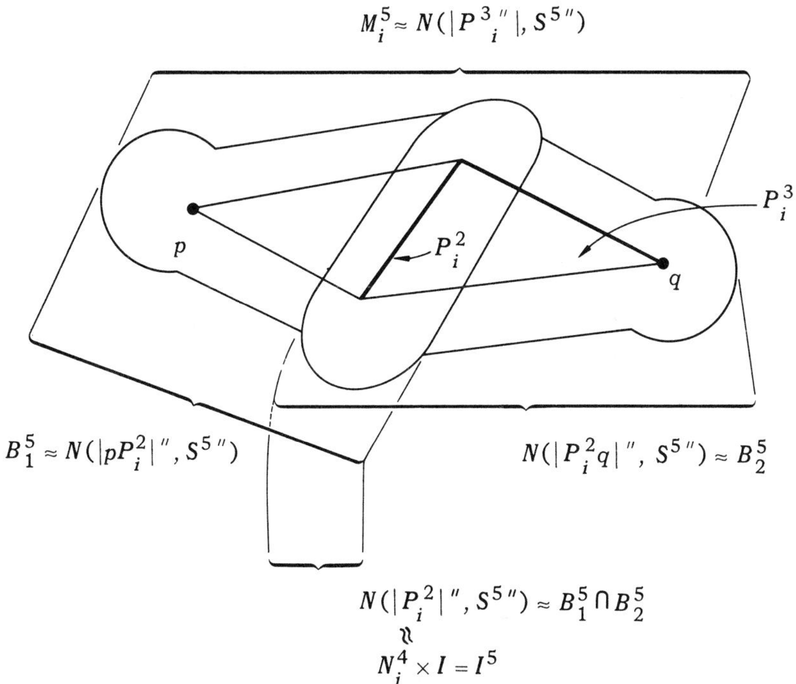

(5) $M_i^5 \times I \approx I^6$ (This will be part (1) of Theorem II.27 for $n=5$). This follows by considering $(\mathrm{Bd} B_1^5) \cap \mathrm{Bd}(N_i^4 \times I) \approx (\mathrm{Bd} N_i^4 \times I) \cup (N_i^4 \times \{1\}) \approx N_i^4$. To show that $M_i^5 \times I \approx I^6$, we will define a pwl homomorphism of $B_2^5 \times I$ onto $M_i^5 \times I$ by taking $(B_1^5 \cap B_2^5) \times I \subset B_2^5 \times I$ onto $B_1^5 \times I$ keeping points fixed in $\mathrm{Cl}((B_2^5 \times I) - [(B_1^5 \cap B_2^5) \times I]) = Z$. We can obtain such a pwl homomorphism since $Z \cap [(B_1^5 \cap B_2^5) \times I] = Z \cap (B_1^5 \times I) = W \approx N_i^4 \times I \approx \hat{B}^5$. Hence $\mathrm{Bd}[(B_1^5 \cap B_2^5) \times I] - \mathrm{int}\, W \approx \hat{B}_1^5$ and $\mathrm{Bd}(B_1^5 \times I) - \mathrm{int}\, W \approx \hat{B}_2^5$ and $\mathring{W} = \dot{\hat{B}}_1^5 = \dot{\hat{B}}_2^5$ and the identity equivalence $\dot{\hat{B}}_1^5 \longrightarrow \dot{\hat{B}}_2^5$ extends to $\hat{B}_1^5 \longrightarrow \hat{B}_2^5$. We now have $\hat{B}_1^5 \cup W = \mathrm{Bd}[(B_1^5 \cap B_2^5) \times I] \longrightarrow \hat{B}_2^5 \cup W = \mathrm{Bd}(B_1^5 \times I)$ and hence this boundary pwl homomorphism extends to a pwl homomorphism $(B_1^5 \cap B_2^5) \times I \longrightarrow B_1^5 \times I$. This homomorphism extends by the identity to Z. Thus

we have the pwl homomorphism $(B_1^5 \cap B_2^5)\times I \cup Z = B_2^5 \times I \longrightarrow B_1^5 \times I \cup Z = M_i^5 \times I$.

THEOREM II.35. *For each $n \geqslant 4$, there exists uncountably many contractible open n-manifolds.*

By Theorems II.26 and II.23 for $n \geqslant 4$ there exists countably many non-homomorphic pseudo n-cells N_i^n $(i=1,2,\ldots)$ such that $N_i^n \times I = I^{n+1}$ for all i. For fixed n, we consider the collection $\{N_i^n\}$ and let $\{M_j\}$ be any infinite subcollection, then the infinite sum $M = M_1 \# M_2 \# M_3 \# \ldots$ is obtained as follows. Let Δ_j, Δ_j' be two disjoint combinatorial $(n\text{-}1)$-cells in $\dot{M}_j$ for every j and $e_j : \Delta_j' \xrightarrow{\text{onto}} \Delta_{j+1}$ be pwl homomorphisms. Then M is obtained from the disjoint union of M_j by identifying each $x \in \Delta_j'$ with $e_j(x) \in \Delta_{j+1}$ for each j. Also the fundamental group of $\dot{M}$ may be written as infinite free product because

$$\pi_1(\mathrm{Bd}(M_1 \# M_2)) \cong \pi_1(M_1) * \pi_1(M_2),$$

$$\pi_1(\mathrm{Bd}(M_1 \# M_2 \# M_3)) \cong \pi_1(M_1) * \pi_1(M_2) * \pi_1(M_3), \text{ etc.}$$

In order to obtain our final two results it will be necessary to quote two results from [20]. To help give more meaning to the first result we want to assume, we will briefly consider some algebraic concepts. Consider a sequence of groups $G_1, G_2, \ldots$ such that

(1) $G_i = A_i^i * A_2^i * \ldots * A_i^i * B_{i+1}^i * B_{i+2}^i * \cdots$

where the A's and B's are arbitrary finitely generated groups and $*$ denotes the free product.

(2) there is a homomorphism $f_i\colon G_{i+1} \longrightarrow G_i$ which

(2.1) maps $A_i^{i+1}, \ldots, A_i^{i+1}$ isomorphically onto $A_i^i, \ldots A_i^i$ respectively,

(2.2) maps A^{i+1}_{i+1} homomorphically into B^i_{i+1} and

(2.3) maps each B^{i+1}_j homomorphically into B^i_j for $j > i+1$.

By defining $f_{ij}: G_i \longrightarrow G_j$ to be the identity if $i = j$ and the composite $f_i f_{i+1} \ldots f_{j-1}$ if $j > i$, we obtain an inverse system $G = \{G_i, f_{ij}\}$ which we call a group system. The abstract group $A^1_1 * A^2_2 * A^3_3 * \ldots$ (infinite free product) will be called the weak limit of the system G and denoted by $\mathrm{WL}(G)$.

Two homomorphisms $f, g: A \longrightarrow B$ are called equal modulo conjugations if there is an inner automorphism h of B such that $f = hg$. This fact will be expressed by $f \equiv g$. Let $H_i = C^i_1 * \ldots * C^i_i * D^i_{i+1} * D^i_{i+2} \ldots$ with finitely generated C's and D's and suppose $H = \{H_i, g_{ij}\}$ is a group system. Let $G = \{G_i, f_{ij}\}$ be a group system as defined above. We say that the systems G and H are compatible if there are integers $i_1 < i_2 < \ldots$ and $j_1 < j_2 < \ldots$ and homomorphisms

$$\ldots \xrightarrow{\lambda_3} G_{i_3} \xrightarrow{\mu_2} H_{j_2} \xrightarrow{\lambda_2} G_{i_2} \xrightarrow{\mu_1} H_{j_1} \xrightarrow{\lambda_3} G_{i_1}$$

such that $\lambda_p \mu_p \equiv f_{i_p i_{p+1}}: G_{i_{p+1}} \longrightarrow G_{i_p}$

and $\mu_p \lambda_{p+1} \equiv g_{j_p j_{p+1}}: H_{j_{p+1}} \longrightarrow H_{j_p}$ for each $p \geqslant 1$.

THEOREM II.36. (Theorem (2.1) of [20]). *Let $G = \{G_i, f_{ij}\}$ and $H = \{H_i, g_{ij}\}$ be group systems. If they are compatible then* $\mathrm{WL}(G) \cong \mathrm{WL}(H)$.

The proof of the above result in [20] assumes for simplicity that each of the A's, B's, C's and D's is non-trivial and indecomposable (i.e., not the free product of two non-trivial groups). This will actually be the case when we want to apply this result.

Since $\mathrm{WL}(G) = A_1^1 * A_2^2 * \ldots$ and $\mathrm{WL}(H) = C_1^1 * C_2^2 * \ldots$ with the A's and B's indecomposable, $\mathrm{WL}(G) \cong \mathrm{WL}(H)$ if and only if for each non-trivial abstract indecomposable group P, there are as many isophormic copies of P among C_j^j, $j = 1, 2, \ldots$ as there are among A , $i = 1, 2, \ldots$. The proof is obtained by showing that for each integer i_1, there exists an integer j_1 such that there are at least as many copies of P among $C_1^{j_1}, C_2^{j_1}, \ldots, C_{j_1}^{j_1}$ as there are among $A_1^{i_1}, A_2^{i_1}, \ldots, A_{i_1}^{i_1}$. This together with its dual will prove the theorem.

THEOREM II.37. (Theorem (4.1) of [20]). *Let M, N be two infinite sums of compact combinatorial n-manifolds with non-empty boundaries such that* int M *and* int N *are homomorphic. Then the corresponding group systems are compatible and in particular,* $\pi_1(\mathring{M}) \cong \pi_1(\dot{N})$.

The corresponding group systems are obtained by making use of the facts that for any $i \geqslant 1$, $\pi_1(\mathrm{Bd}(M_1 \# M_2 \# \ldots \# M_i)) \cong \pi_1(\dot{M}_1) * \pi_1(\dot{M}_2) * \ldots * \pi_1(\dot{M}_i)$; that for any $i \geqslant 1$ $\mathrm{Bd}(M_1 \# M_2 \# \ldots \# M_i)$ has a collar, say $\mathcal{C}_i$, in $M_1 \# M_2 \# \ldots \# M_i$; and that for all $i \geqslant 1$ we can obtain these collars so that $U_i = \mathcal{C}_i \cup M_{i+1} \# M_{i+2} \# \ldots$ contains $U_{i+1} = \mathcal{C}_{i+1} \cup M_{i+2} \# M_{i+3} \# \ldots$ and $\cap U_i = \mathrm{Bd}(M_1 \# M_2 \# \ldots)$. The corresponding group systems are shown to be compatible because there is a homomorphism h: int $M \longrightarrow$ int N and π_1 of a path connected space is invariant up isomorphism by path conjugation under changes of base points.

Proof of Theorem II.35. By Theorem II.23 and its proof for each $n \geqslant 5$ there exists a collection $\{N_i^n\}$ of compact combinatorial n-manifolds with non-empty boundaries such that $\pi_1(\mathrm{Bd}\,N_i^n)$ is a non-trivial indecomposable group ($\pi_1(\mathrm{Bd}\,N_i^n)$ is a perfect group and has a presentation consisting of two generators and two relations) and for $i \neq j$ $\pi_1(\mathrm{Bd}N_i^n) \neq \pi_1(\mathrm{Bd}N_j^n)$. By Theorem II.26 for $n \geqslant 4$ we obtain a similar collection $\{N_i^n\}$. $\pi_1(\mathrm{Bd}N_i^n)$ is a non-trivial perfect group and for $i \neq j$ $\pi_1(\mathrm{Bd}N_i^n) \neq \pi_1(\mathrm{Bd}N_j^n)$. If we use the collection $\{W_i^n\}$ instead then we also have that $\pi_1(\mathrm{Bd}W_i^n)$ is a non-trivial indecomposable group (it also is perfect and has a presentation consisting of two generators and two relations). Thus Theorem II.35 follows by forming infinite sums of M_i's for the collection $\{N_i^n\}$, for fixed n, in uncountably different ways such that in two different ones some M_i occurs more in one than in the other. The desired contractible open n-manifolds are then merely the interiors of the uncountable collection of infinite sums. It follows by Theorem II.37 that they all are distinct.

THEOREM II.38. *For $n \geqslant 4$, E^n has uncountably many different involutions, any two distinguished by the fundamental groups of their fixed-point sets.*

Proof. Each M_j in the infinite sums constructed has the property that $M_j \times I = I^{n+1}$ and $S^n = \mathrm{Bd}(M_j \times I) = 2M_j$. It follows then that for any i, $(M_1 \# M_2 \# \ldots \# M_i) \times I = I^{n+1}$ and that $S^n = 2(M_1 \# M_2 \# \ldots \# M_i)$. Let B^n be the unit n-ball in E^n and $p = (0, \ldots, 0, 1) \in E^{n-1}\text{-}E^n$. Let $D^{n+1} = pB^n$ be the n-cell obtained by forming the cone over B^n from p. There is a natural homeomorphism $g\colon B^n \times [0,1) \longrightarrow pB^n - \{p\}$ so that $g(b,0) = b \in B^n \subset pB^n$. Given an infinite sum $M = M_1 \# M_2 \# \ldots$ we can define by induction a homeomorphism $h\colon M \times I \longrightarrow pB^n - \{p\}$ by sending $M_j \times I \subset M \times I$ homeomorphically to $g(B^n \times [1-(1/j+1), 1-(1/j)])\subset pB^n - \{p\}$ so that $M_{j-1} \cap M_j \times I \subset M \times I$ goes to $g(B^n \times (1-(1/j)))$ and $(M_j \cap M_{j+1}) \times I \subset M \times I$ goes to $g(B^n \times (1-(1/j+1)))$. Thus $M \times I$ is homeomorphic to $pB^n - \{p\}$ and $2(M_1 \# M_2 \# \ldots) = \mathrm{Bd}(M \times I)$ is homeomorphic to

$\text{Bd}(pB^n)$-$\{p\}$ which is homeomorphic to E^n. Thus for each infinite sum $M = M_1 \# M_2 \# \ldots$ we have that $2(M_1 \# M_2 \# \ldots)$ is homeomorphic to E^n and we obtain an involution of E^n by interchanging the copies of $M_1 \# M_2 \# \ldots$ in $2(M_1 \# M_2 \# \ldots)$ leaving $\text{Bd}(M_1 \# M_2 \# \ldots)$ fixed and the result follows.

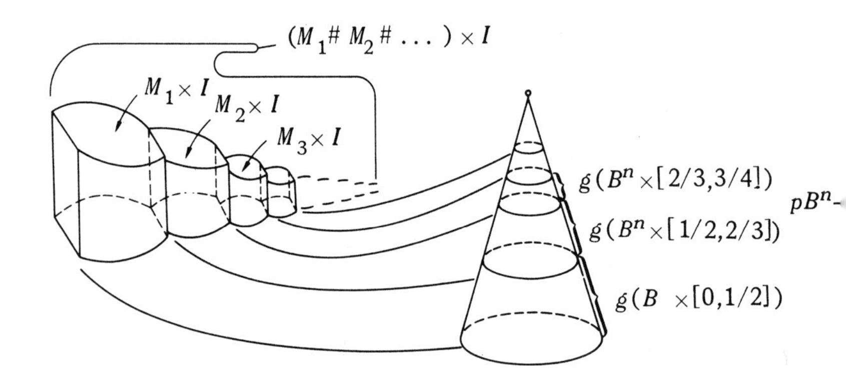

§ C. Embedding of Manifolds in Euclidean Space

The ideas of this section come from the paper by R. Penrose, J.H.C. Whitehead, and E.C. Zeeman, 'Imbedding of Manifolds in Euclidean Space', *Annals of Math.* 73 (1961), 613–623. The main results are:

THEOREM II.39. *If* $0<2m\leqslant n$, *then every closed combinatorial* $(m\text{-}1)$-*connected* n-*manifold can be combinatorially embedded in* E^{2n-m+1}.

THEOREM II.40. *Let* M *be a compact, bounded combinatorial* n-*manifold which is* $(m\text{-}1)$-*connected* $(0<2m\leqslant n)$. *If either*

(a) $\mathring{M}\times I$ *can be embedded in* E^{2n-m}, *or*

(b) $\dot{M}$ *is* $(m\text{-}2)$-*connected* ((-1)-*connected means non-vacuous*), *then* M *can be combinatorially embedded in* E^{2n-m}.

Since the regular neighborhood of $\mathring{M}$ in M is $\approx \mathring{M}\times I$, condition (a) is certainly necessary for the embeddability of M in E^{2n-m}. This is obviously satisfied if $\dot{M}$ can be embedded in E^{2n-m-1}. To obtain the above results we first prove some lemmas.

Definitions. A map $f\colon X \longrightarrow Y$, where X, Y are arbitrary topological spaces, is called an embedding if and only if, it is a homomorphism onto $f(X)$. A map $f\colon X \longrightarrow Y$ is called an immersion if and only if every point $x\in X$ has a neighborhood $N_x\subset X$ such that $f|_{N_x}$ is an embedding.

LEMMA II.41. *Let $f\colon X \longrightarrow Y$ be an immersion of a locally compact metric space X in a Haudorff space Y and let $f|A$ be an embedding, where A is a compact subset of X. Then there is a compact neighborhood $N \subset X$ of A such that $f|N$ is an embedding.*

Proof. Let $N_i = \{x \in X | \rho(x, A) \leqslant 1/i\}$ $(i = 1, 2, \ldots)$, where ρ is a metric for X. Then there is an integer k such that N_i is compact if $i \geqslant k$. Assume that $f|N_i$ is not an embedding for any i. Then for each $i > k$, there are points $x_i, x_i' \in N_i$ such that $x_i \neq x_i'$, but $f(x_i) = f(x_i') = y$. This follows since N_i is compact for $i \geqslant k$ and Y is a Hausdorff space. Since f is locally 1-1 and $f|A$ is 1-1 it follows that some subsequence of the sequence $\{y_i\}$ converges to each of two distinct points in $f(A)$, which is impossible.

Definition. Let P, Q be compact polyhedra in a combinatorial manifold M. We say Q is quasi-complementary to P, if and only if every compact polyhedron in M-P can be piecewise linearly embedded in every neighborhood of O. For example, let K be a triangulation of M and let the vertices K be separated into two disjoint subsets A, B. Let P be the union of the simplexes of K whose vertices are all in A and Q the union of the simplexes of K whose vertices are all in B. As another example suppose M is a compact manifold and T is a triangulation of M. Let K be an arbitrary proper subcomplex of T and L and the subcomplex of T' maximal with respect to missing K, then each of K' and L are quasi-complementary to each other. If M is a combinatorial n-sphere, then any proper, non-empty, compact polyhedral subsets of M are quasi-complementary to each other.

LEMMA II.42. *Let M be a compact (possibly bounded) $(m-1)$-connected combinatorial n-manifold, where $0 < 2m \leqslant n$.*

Assume that there are compact polyhedra $P \subset \operatorname{int} M$, $Q \subset M$ such that $\dim P < m$, Q is quasi-complementary to P and some neighborhood of Q can be combinatorially embedded in E^q. Then M can be combinatorially embedded in E^{q+1} and in E^q if it is bounded.

Proof. Let $U \subset M$ be a neighborhood of Q which can be embedded in E^q. Since $P \subset \operatorname{int} M$, $\dim P < m$ and M is $(m-1)$-connected it follows from Corollary IV.15 in Volume I, that if M is bounded, it can be embedded in $M-P$, hence in U and hence in E^q.

Now let M be unbounded. Then by Proposition IV.14, there is a combinatorial n-ball $E \subset M$ such that $P \subset \operatorname{int} E$. Let $M_0 = M - \operatorname{int} E$. Then M_0 can be embedded in U and there is therefore an embedding $f\colon M_0 \longrightarrow E^q$. We take E^q to be a hyperplane in E^{q+1} and extend f to an embedding $M \longrightarrow E^{q+1}$, which maps E on the join of $f(\dot{E})$ and a point in $E^{q+1} - E^q$.

LEMMA II.43. *Let Q be a compact polyhedron in a combinatorial manifold M and let Q have a neighborhood which can be immersed in E^q, where $q > 2 \dim Q$. Then O has a neighborhood which can be embedded in E^q.*

Proof. This follows from the properties of general position and from Lemma II.41.

Definition. By the branch locus of a pwl map $f\colon M \longrightarrow E^q$ we mean the set of points $x \in M$ such that no neighborhood of x is embedded by f. Let K be a rectilinear triangulation of M such that f is barycentric in each simplex of K (we do not assume that f is simplicial with respect to K and a triangulation of E^q). Then the branch locus, B, of f is the union of all closed simplexes $\sigma \in K$ such that $f | \operatorname{st}(\sigma, K)$ is not an embedding. Hence B is a compact polyhedron in M and $f | M - B$ is an immersion.

LEMMA II.44. *Let M be a closed $(m-1)$-connected combinatorial n-manifold, where $0<2m\leqslant n$. Assume that there is a pwl map $M \longrightarrow E^q$ whose branch locus is at most $(m-1)$-dimensional, where $q>2(n-m)$. Then M can be emdded in E^{q+1}.*

Proof. Let $f: M \longrightarrow E^q$ be a map whose branch locus is at most $(m-1)$-dimensional and let K be a triangulation of M such that f is barycentric in each simplex of K. Let B be the $(m-1)$-skeleton of K and let L be the subcomplex of K' maximal with respect to missing B. Then $f|M-B$ is an immersion and $\dim L = n-m$. Since $q>2(n-m)$, taking $|L^{n-m}|$ to be the Q of Lemma II.43, then L has a neighborhood which can be embedded in E^q. Now taking $P = |B'|$ and $Q = |L^{n-m}|$ in Lemma II.42, we get that M can be embedded in E^{q+1}.

Now we can return to a proof of Theorem II.39. Let $f: M \longrightarrow E^{2n-m}$ be a map which is barycentric in each simplex of a triangulation K of M and which maps the vertices of K in general position. Now f embeds each simplex of K. Let σ_1, σ_2 be simplexes of K and let $\sigma_0 = \sigma_1 \cap \sigma_2$ and suppose $\dim \sigma_0 = p \geqslant m$, $\dim \sigma_1 = r$ and $\dim \sigma_2 = s$. Then σ_1 is the join of σ_0 and an $(r-p-1)$-simplex τ. Since $(r-p-1) + s - (2n-m) \leqslant m-p-1<0$ and the vertices of K are mapped in general position, it follows that $f(\tau)$ does not meet the s-plane containing $f(\sigma_2)$. Thus $f(\sigma_1) \cap f(\sigma_2) = f(\sigma_0)$ and hence $f|\mathrm{st}(\sigma, K)$ is an embedding if $\dim \sigma \geqslant m$ $(\sigma \in K)$. Therefore, the branch locus of f is at most $(m-1)$-dimensional and the result follows from Lemma II.44 since $2n-m>2(n-m)$.

Proof of Theorem II.40. Let $\dot{M} \times I$ be embeddable in E^{2n-m}. Then there is a closed polyhedral neighborhood $N \subset M$ of $\dot{M}$ which is pwl homomorphic to $M \times I$ and hence embeddable in E^{2n-m}. Let K be the triangulation of the pair (M, N) such that the given embedding $N \longrightarrow E^{2n-m}$ is barycentric in each simplex of $\dot{K} \cap N$ and so that $\dot{K}$ is complete (full) in K. Extend this to a map $f\colon K \longrightarrow E^{2n-m}$ which is barycentric in each simplex of K and maps the vertices of K in general position. Let P denote the union of all simplexes of K which are at most $(m-1)$-dimensional and do not meet $\dot{M}$. Then the branch locus of f is contained in P.

Let K' denote the first barycentric subdivision of K and $K_0 = K' - 0(\dot{K}', K')$, where $0(\dot{K}', K')$ denotes the union of the open simplexes of K' whose closures meet $\dot{K}'$. From results developed earlier we know that $M_0 = |K_0| \approx M$. Hence it is enough to prove that M_0 can be embedded in E^{2n-m}.

We say that a vertex $\hat{\sigma} \in K_0$ ($\hat{\sigma}$ is the barycenter of $\sigma \in K$) is of the first kind if dim $(\sigma) < m$ and $\sigma \cap \dot{K} = \emptyset$. $\hat{\sigma}$ is of the second kind if dim $(\sigma) \geqslant m$ or $\sigma \cap \dot{K} \neq \emptyset$. Then the polyhedron P is the union of the simplexes of K_0 whose vertices are all of the first kind. Let Q denote the union of the simplexes of K_0 whose vertices are all of the second kind. Then Q is quasi-complementary to P. It consists of all simplexes $\hat{\sigma}_0 \ldots \hat{\sigma}_p \in K_0$ such that either dim $(\sigma_0) \geqslant m$, hence $p \leqslant n-m$, or $\sigma_0 \cap \dot{K} \neq \emptyset$, which means that $\hat{\sigma}_0 \ldots \hat{\sigma}_p \in \dot{K}_0$. Also $Q \subset M_0 - P$ and $\dot{M}_0 \subset N$, hence $f|\dot{M}_0$ is an embedding.

Let the images of the vertices of K_0 be shifted slightly so as to define a map $f_0\colon M_0 \longrightarrow E^{2n-m}$, barycentric in each simplex of K_0, such that $f_0|\dot{M}_0$ is an embedding, $f_0|M_0 - P$ is an immersion and f_0 maps the vertices of K_0 in general position. Since dim $(Q - \dot{M}_0) \leqslant n-m$ and $(n-m)+(n-1) < 2n-m$ it follows that $f_0|Q$ is an embedding. By Lemma II.41, f_0 embeds some neighborhood of Q and by Lemma II.42 the result follows for case (a).

Now let $\dot{M}$ be $(m-2)$-connected and we consider $2M = M \cup M_1$, where $\dot{M} = \dot{M}_1 = M \cap M_1$ and M_1 is another copy of M. Then using van Kampen's Theorem and the Mayer-Vietoris sequence, if $m > 1$, we get that $2M$ is $(m-1)$-connected. Let E be an n-ball in M_1 and let $M_2 = 2M - \text{int } E$. Then M_2 is $(m-1)$-connected and $\dot{M}_2$ is the $(n-1)$-sphere $\dot{E}$. Then since $\dot{E} \times I$ can be embedded in E^{2n-m} it follows from case (a) that M_2 can be embedded in E^{2n-m}. Since $M \subset M_2$, this completes the proof.

§ D. Taming Complexes

The ideas discussed here are contained in the papers by Tatsuo Homma, 'On Imbedding of Polyhedra in Manifolds' and by Herman Gluck, 'Embeddings in the Trivial Range', *Bulletin of the A.M.S.* 69 (1963), 824-831. The main results are as follows.

THEOREM II.45 (Homma). *Let M^n, $\tilde{M}^n$ and $\tilde{P}^k$ be two finite combinatorial n-manifolds (possible with boundary) and a finite simplicial complex such that $\tilde{M}^n$ is topologically embedded in M^n, $\tilde{P}^k$ is piecewise linearly embedded in* int $(\tilde{M}^n)$ *and $2k+2 \leqslant n$. Then for any $\varepsilon > 0$ there is an ε-homomorphism h of M^n onto itself such that $h | M^n - U_\varepsilon(\tilde{P}^k) =$* identity *and $h | \tilde{P}^k : \tilde{P}^k \longrightarrow M^n$ is piecewise linear.*

THEOREM II.46. (Gluck's modification of Homma' Theorem): *Let the following be given:*

(1) *M^n, a possibly noncompact combinatorial n-manifold,*

(2) *$\tilde{M}^n$, a possibly noncompact combinatorial n-manifold, topologically embedded in M^n,*

(3) *$\tilde{P}^k$, a possible infinite (locally finite) simplicial complex, pwl embedded as a closed subset of $\tilde{M}^n$,*

(4) *$\tilde{L}$, a subcomplex of $\tilde{P}^k$ such that the closure of $\tilde{P}^k - \tilde{L}$ is a finite complex, and such that $\tilde{L}$ is pwl embedded in M^n as well as in $\tilde{M}^n$.*

If $2k+2 \leqslant n$, then for any $\varepsilon > 0$ there is an ε-push h of $(M^n, \tilde{P}^k - \tilde{L})$ such that $h | \tilde{P}^k : \tilde{P}^k \longrightarrow M^n$ is pwl and $h | \tilde{L} =$ identity.

Before we state the remaining main results we will first make some definitions. Let $f\colon P^k \longrightarrow M^n$ be an embedding. If there is a triangulation of P^k such that for each point $x \in P^k$ there is an open neighborhood U of $f(x)$ in M^n and a triangulation of U as a combinatorial manifold in terms of which f is pwl on some neighborhood of x, then we will say that f is a locally tame embedding. Note that the triangulation of U need not be related to any triangulation of M^n. Also once a triangulation is chosen for P^k, the same triangulation must be used for deciding whether f is locally tame at each point $x \in P^k$. However, if another embedding f' is given, an entirely different triangulation of P^k may be used to decide whether f' is locally tame.

Let M be a topological manifold with a metric d, and A a subset of M. If $\varepsilon > 0$, an ε-push of (M, A) is a homomorphism of M onto itself such that

(1) h is an ε-homomorphism (i.e., $d(x, h(x)) < \varepsilon$ for all $x \in M$)

(2) $h | M - U_\varepsilon(A) =$ identity (i.e., if $d(x, A) \geqslant \varepsilon$, then $h(x) = x$. $U_\varepsilon(A) = \{x \in M \mid d(x, A) < \varepsilon\}$),

(3) h is ε-isotopic to the identity under an isotopy which restricts to the identity on $M - U_\varepsilon(A)$ (i.e., each h of the isotopy satisfies conditions (1) and (2) above).

THEOREM II.47. *Let f be a locally tame embedding of the polyhedron P^k into the combinatorial manifold M^n. If $2k+2 \leqslant n$, then for each $\varepsilon > 0$ there is an ε-push of $(M^n, f(P^k))$ such that $hf\colon P^k \longrightarrow M^n$ is pwl with respect to arbitrary preassigned triangulations of P^k and M^n.*

THEOREM II.48. *If $2k+2 \leqslant n$, then for each $\varepsilon > 0$ there is a $\delta > 0$ such that if f and f' are any two locally tame embeddings of the polyhedron P^k into the combinatorial manifold M^n with $d(f, f') < \delta$,* *then there exists an ε-push h of $(M^n, f(P^k))$ satisfying $hf = f'$.*

THEOREM II.49. *Let f and f' be locally tame embeddings of the polyhedron P^k into the combinatorial manifold M^n. If $2k+2 \leqslant n$ and f is homotopic to f', then there is a homomorphism h of M^n onto itself which is isotopic to the identity, such that $hf = f'$.*

We will first obtain some lemmas necessary to prove Theorem II.46; clearly Theorem II.45 is just a special case of Theorem II.46. Also for convenience we will restate Theorem IV.31 from Volume I, with an easy generalization, in a form in which we will want to apply it.

PROPOSITION II.50. *Let P^k be a (possible infinite) k-dimensional polyhedron pwl embedded as a closed subset of E^n, $2k+2 \leqslant n$, and L a subcomplex of P^k such that the closure of $P^k - L$ is a finite complex, and suppose f is a pwl homomorphism of P^k into E^n such $f(x) = x$ for $x \in L$ and f does not move any other point as far as ε. Then there is an isotopy h_t of E^n onto itself such that*

(1) *h_0 = identity*

(2) *$h_1 = f$ on P^k*

(3) *each h_t is the identity outside an ε-neighborhood of $\overline{P^k - L}$*

(4) *each point of E^n moves along a polygonal path of length less than ε.*

PROPOSITION II.51. *Theorem II.46 with $\tilde{M}^n = \tilde{E}^n$.*

To prove this proposition we are going to construct two sequences of homomorphisms $\{K_i\}$, $\{H_i\}$, where each homomorphism carries M^n onto itself and we want $\lim_{i\to\infty} H_i = H$ and $\lim_{i\to\infty} K_i = K$ also to be homomorphisms carrying M^n onto itself. For this purpose we quote an elementary lemma often used in point-set topology.

Lemma for Proposition II.51. *Suppose $F_1, F_2, \ldots$ is a sequence of homomorphism of a metric space X into a complete metric space Y such that $F_i = F_j$ outside some fixed compact subset of X for all i and j, and $\varepsilon_1, \varepsilon_2, \ldots$ is a sequence of positive numbers such that $d(F_i(x), F_{i+1}(x)) < \varepsilon_i$ for all $x \in X$. Then $\lim_{i\to\infty} F_i(x)$ is a homomorphism if $\Sigma_{i=1}^{\infty} \varepsilon_i$ is a positive number so small that if $d(x_1, x_2) > 1/j$ $(x_1, x_2 \in X)$, then $d(F_j(x_1), F_j(x_2)) > 2\,\Sigma_{i=j}^{\infty} \varepsilon_i$.*

Proof of Proposition II.51. Given $\varepsilon > 0$, let $\varepsilon' = d(\overline{\tilde{P}-\tilde{L}}, \tilde{M}^n - \tilde{E}^n)$ and take $\tilde{\varepsilon} = \min(\varepsilon, \varepsilon', t_0)$ where $0 < t_0 < 1$. Let P be a polyhedron in M^n approximating $\tilde{P}^k$ (obtained by putting the vertices of $\tilde{P}^k - \tilde{L}$ in general position in $|\tilde{E}^n| \cap M^n$ under the combinatorial triangulation of M^n) so that under the induced equivalence $\tilde{P} \approx P$, the pwl homomorphism φ: $\tilde{P} \longrightarrow P$ is such that $d(\varphi, \text{identity}) < \tilde{\varepsilon}/8$ (i.e., for $x \in \tilde{P}$, $d(\varphi(x), x) < \tilde{\varepsilon}/8$ and of course $\varphi(x) = x$ for $x \in \tilde{L}$ since $\tilde{L}$ is pwl embedded in M^n as well as in $\tilde{E}^n$).

Let $\tilde{P}_1$ be an approximation to P under $\tilde{E}^n$ so that under the pwl homomorphism $\tilde{f}_1 : \tilde{P} \longrightarrow \tilde{P}_1$ we have $d(\tilde{f}_1, \varphi) < \tilde{\varepsilon}/8$ and hence $d(\tilde{f}_1, \text{identity}) < \tilde{\varepsilon}/4$. Let $\tilde{\varepsilon}_1 = \tilde{\varepsilon}/4$. As above $\tilde{f}_1 | \tilde{L} = \text{identity}$ and we will leave $\tilde{L}$ fixed under each additional approximation. By Proposition II.50 $\tilde{f}_1$ extends to $\tilde{h}_1$: $\tilde{E}^n \longrightarrow \tilde{E}^n$ so that $d(\tilde{h}_1, \text{identity}) < \tilde{\varepsilon}_1$ and $\tilde{h}_i = \text{identity}$ on $\tilde{E}^n - U_{\tilde{\varepsilon}_1}(\mathrm{Cl}(\tilde{P}^k - \tilde{L}))$. In fact there is an isotopy connecting the identity homomorphism and $\tilde{h}_1$ with similar properties. $\tilde{h}_1$ (and the isotopy) extends to all of M^n by defining it to be the identity outside $\tilde{E}^n$. In each case we make a new approximation, we will use Proposition II.50 to extend the given approximation homomorphism to all of M^n and we will have an isotopy between the identity and the new extended homomorphism. So for convenience we will not mention the isotopy during the various stages of construction, but will just note here that there is one each time.

Let us denote by E^n the subpolyhedron of $|\tilde{E}^n| \cap M^n$. That is the combinatorial triangulation we take for E^n will be compatible with that given on M^n. In any event we will only be moving points in a neighborhood of $\mathrm{Cl}(P^k - \tilde{L}) \subset E^n$. Let $\tilde{\sigma}_1 = \text{g.l.b.}\ \{d(\tilde{h}_1(x), \tilde{h}_1(x')) |$ where $d(x, x') \geqslant \frac{1}{2}\}$.

Let P_1 be an approximation to $\tilde{P}_1$ under E^n so that under the induced $g_1 \colon P \longrightarrow P_1$, $d(g_1, \tilde{f}_1 \circ \varphi^{-1}) < \min(\tilde{\varepsilon}_1/4, \tilde{\varepsilon}/8, \tilde{\sigma}_1/16)$. Hence $d(g_1, \text{identity}) < \tilde{\varepsilon}/8 + \tilde{\varepsilon}/8 = \tilde{\varepsilon}/4$. Let $\varepsilon_1 = \tilde{\varepsilon}/8$. As above g_1 extends to k_1 on all of M^n so that $d(k_1, \text{identity}) < \varepsilon_1$. Let $\sigma_1 = \text{g.l.b.}\ \{d(k_1(x), k_1(x')) \mid d(x, x') \geqslant \frac{1}{2}\}$.

Let $\tilde{P}_2$ approximate P_1 under $\tilde{E}^n$ and $\tilde{f}_1 \colon \tilde{P}_1 \longrightarrow \tilde{P}_2$ so that $d(\tilde{f}_2 \circ \tilde{f}_1, g_1 \circ \varphi) < \min(\tilde{\varepsilon}_1/4, \varepsilon_1/4, \tilde{\sigma}_1/16, \sigma_1/16)$. Hence $d(\tilde{f}_2, \text{identity}) < \tilde{\varepsilon}_1/4 + \tilde{\varepsilon}_1/4 = \tilde{\varepsilon}_1/2$ and $< \tilde{\sigma}_1/16 + \tilde{\sigma}_1/16 = \tilde{\sigma}_1/8$. $\tilde{f}_2$ extends to $\tilde{h}_2 \colon M^n \longrightarrow M^n$ so that $d(\tilde{h}_2, \text{identity}) < \tilde{\varepsilon}_1/2$ and $< \tilde{\sigma}_1/8$. Let $\tilde{\varepsilon}_2 = \min(\tilde{\varepsilon}_1/2, \tilde{\sigma}_1/8)$.

Let $H_2 = \tilde{h}_2 \circ \tilde{h}_1$ (the K_1 and H_1 mentioned in the remarks above are merely k_1 and h_1 respectively). Let $\tilde{\sigma}_1 = \text{g.l.b.}\ \{d(H_2(x), H_2(x)) \mid d(x, x') \geqslant \frac{1}{3}\}$.

Let P_2 approximate $\tilde{P}_2$ under E^n and $g_2 \colon P_1 \longrightarrow P_2$ so that $d(g_2 \circ g_1, \tilde{f}_2 \circ \tilde{f}_1 \circ \varphi^{-1}) < \min(\tilde{\varepsilon}_2/4, \varepsilon_1/4, \tilde{\sigma}_2/16, \sigma_1/16)$. Hence $d(g_2, \text{identity}) < \varepsilon_1/2$ and $\sigma_1/8$. As before g_2 extends to $k_2 \colon M^n \longrightarrow M^n$ and $d(k_2, \text{identity}) < \varepsilon_1/2$ and $< \sigma_1/8$. Let $\varepsilon_2 = \min(\varepsilon_1/2, \sigma_1/8)$. Let $K_2 = k_2 \circ k_1$ and $\sigma_2 = \text{g.l.b.}$ $\{d(K_2(x), K_2(x')) \mid d(x, x') \geqslant \frac{1}{3}\}$.

In general we have: $\tilde{P}_i$ approximates P_{i-1}, $\tilde{f}_i \colon \tilde{P}_{i-1} \longrightarrow \tilde{P}_i$, $d(\tilde{f}_i \circ \ldots \circ \tilde{f}_1, g_{i-1} \circ \ldots \circ g_1 \circ \varphi) < \min(\tilde{\varepsilon}_{i-1}/4, \varepsilon_{i-1}/4, \tilde{\sigma}_{i-1}/16, \sigma_{i-1}/16)$, and $d(\tilde{f}_i$, identity) $< \tilde{\varepsilon}_{i-1}/2$ and $\tilde{\sigma}_{i-1}/8$. Now $\tilde{f}_i$ extends to $\tilde{h}_i$: $M^n \longrightarrow M^n$ and $d(\tilde{h}_i$, identity) $< \tilde{\varepsilon}_{i-1}/2$ and $\tilde{\sigma}_{i-1}/8$. Let $H_i = \tilde{h}_i \circ H_{i-1}$ and $\tilde{\sigma}_i = \text{g.l.b.} \{d(H_i(x), H_i(x')) \mid d(x, x') \geqslant 1/_{i+1}\}$. Also we let $\tilde{\varepsilon}_i = \min(\tilde{\varepsilon}_{i-1}/2, \tilde{\sigma}_{i-1}/8)$.

P_i approximates $\tilde{P}_i$, g_i: $P_{i-1} \longrightarrow P_i$, $d(g_i \circ \ldots \circ g_1, \tilde{f}_i \circ \ldots \tilde{f}_1 \circ \varphi^{-1}) < \min(\tilde{\varepsilon}_i/4, \varepsilon_{i-1}/4, \tilde{\sigma}_i/16, \sigma_{i-1}/16)$ and $d(g_i$, identity) $< \varepsilon_{i-1}/2$ and $< \sigma_{i-1}/8$. Now g_i extends to k_i: $M^n \longrightarrow M^n$ and $d(k_i$, identity) $< \varepsilon_{i-1}/2$ and $< \sigma_{i-1}/8$. Let $K_i = k_i \circ K_{i-1}$ and $\sigma_i = \text{g.l.b.} \{d(K_i(x), K_i(x')) \mid d(x, x') \geqslant 1/_{i+1}\}$. Also we let $\varepsilon_i = \min(\varepsilon_{i-1}/2, \sigma_{i-1}/8)$.

The following diagram illustrates the above facts where the dotted arrows point to the new approximating polyhedra; the vertical arrows indicate the pwl homomorphisms; and the numbers on or along side the arrows give an upper bound on the distance between the images of points under the appropriate mappings of the sets connected by the arrows.

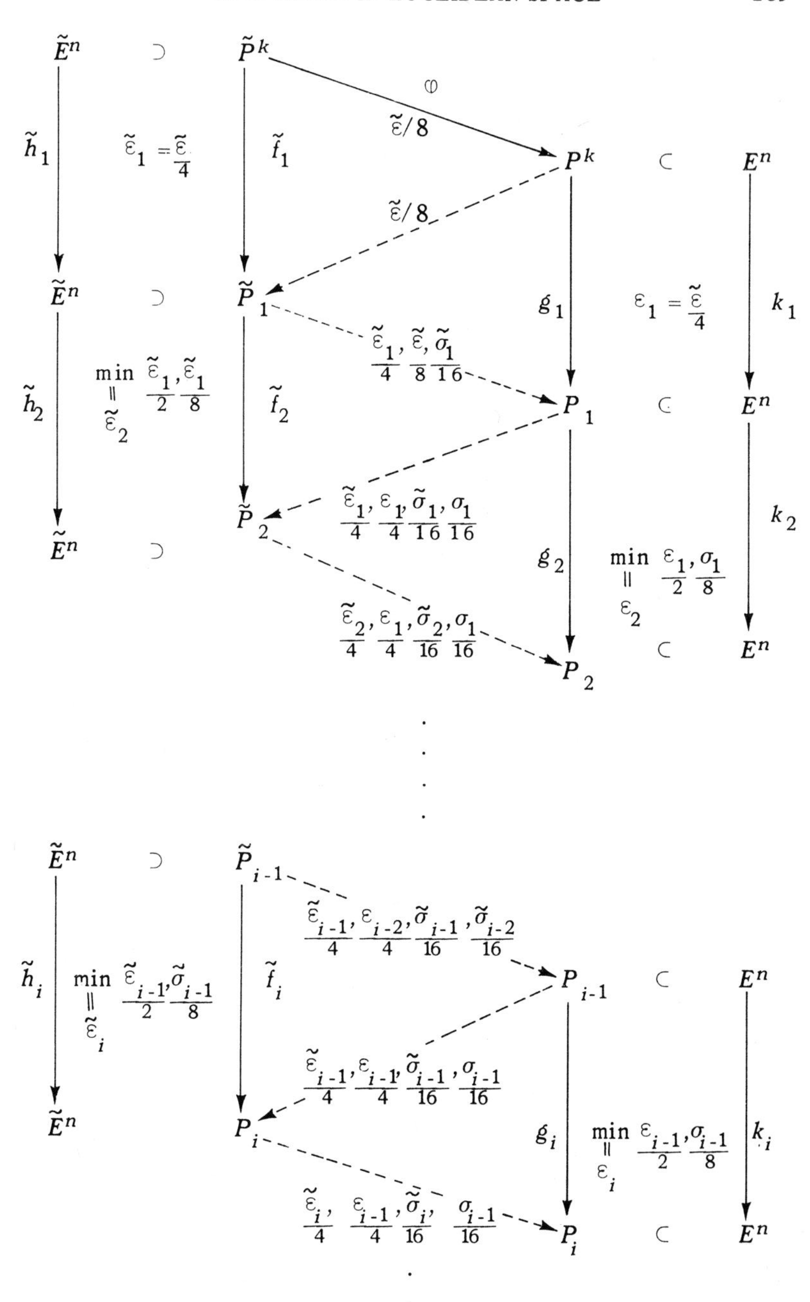

$\tilde{E}^n \supset \tilde{P}^k$
φ
$\tilde{\varepsilon}/8$
$\tilde{h}_1$
$\tilde{\varepsilon}_1 = \frac{\tilde{\varepsilon}}{4}$
$\tilde{f}_1$
$P^k \subset E^n$
$\tilde{\varepsilon}/8$
$\tilde{E}^n \supset \tilde{P}_1$
g_1
$\varepsilon_1 = \frac{\tilde{\varepsilon}}{4}$
k_1
$\frac{\tilde{\varepsilon}_1}{4}, \frac{\tilde{\varepsilon}}{8}, \frac{\tilde{\sigma}_1}{16}$
$\tilde{h}_2$
$\min \frac{\tilde{\varepsilon}_1}{2}, \frac{\tilde{\varepsilon}_1}{8} = \tilde{\varepsilon}_2$
$\tilde{f}_2$
$P_1 \subset E^n$
$\frac{\tilde{\varepsilon}_1}{4}, \frac{\varepsilon_1}{4}, \frac{\tilde{\sigma}_1}{16}, \frac{\sigma_1}{16}$
$\tilde{P}_2$
k_2
$\tilde{E}^n \supset$
g_2
$\min \frac{\varepsilon_1}{2}, \frac{\sigma_1}{8} = \varepsilon_2$
$\frac{\tilde{\varepsilon}_2}{4}, \frac{\varepsilon_1}{4}, \frac{\tilde{\sigma}_2}{16}, \frac{\sigma_1}{16}$
$P_2 \subset E^n$
$\tilde{E}^n \supset \tilde{P}_{i-1}$
$\frac{\tilde{\varepsilon}_{i-1}}{4}, \frac{\varepsilon_{i-2}}{4}, \frac{\tilde{\sigma}_{i-1}}{16}, \frac{\tilde{\sigma}_{i-2}}{16}$
$\tilde{h}_i$
$\min \frac{\tilde{\varepsilon}_{i-1}}{2}, \frac{\tilde{\sigma}_{i-1}}{8} = \tilde{\varepsilon}_i$
$\tilde{f}_i$
$P_{i-1} \subset E^n$
$\frac{\tilde{\varepsilon}_{i-1}}{4}, \frac{\varepsilon_{i-1}}{4}, \frac{\tilde{\sigma}_{i-1}}{16}, \frac{\sigma_{i-1}}{16}$
$\tilde{E}^n$
P_i
g_i
$\min \frac{\varepsilon_{i-1}}{2}, \frac{\sigma_{i-1}}{8} = \varepsilon_i$
k_i
$\frac{\tilde{\varepsilon}_i}{4}, \frac{\varepsilon_{i-1}}{4}, \frac{\tilde{\sigma}_i}{16}, \frac{\sigma_{i-1}}{16}$
$P_i \subset E^n$

Inductively then, we obtain two sequences of homomorphisms $\{H_i\}$, $\{K_i\}$ such that $d(H_i, H_{i+1}) < \tilde{\varepsilon}_{i+1} \leqslant \tilde{\varepsilon}_i/2$ and $\tilde{\sigma}_i/8$ (also $< \tilde{\varepsilon}/2^{i+1}$) and $d(K_i, K_{i+1}) < \varepsilon_{i+1} \leqslant \varepsilon_i/2$ and $\sigma_i/8$ (also $< \tilde{\varepsilon}/2^{i+1}$). If $j \geqslant 2$ and $d(x_1, x_2) > 1/j$ then $d(K_{j-1}(x_1), K_{j-1}(x_2)) > \sigma_{j-1}$ and $d(H_{j-1}(x_1), H_{j-1}(x_2)) > \tilde{\sigma}_{j-1}$. Since $d(\tilde{h}_j$, identity$) < \tilde{\sigma}_{j-1}/8$ and $d(k_j$, identity$) < \sigma_{j-1}/8$ it follows that $d(K_j(x_1), K_j(x_2)) > 3\sigma_{j-1}/4$ and $d(H_j(x_1), H_j(x_2)) > 3\tilde{\sigma}_{j-1}/4$. Now $2\sum_{i=j}^{\infty} \tilde{\varepsilon}_i \leqslant 2\sum_{i=j}^{\infty} (\tilde{\sigma}_{j-1}/8)(1/2_{i-1}) = \tilde{\sigma}_{j-1}/2$ and similarly $2\sum_{i=1}^{\infty} \varepsilon_i \leqslant \sigma_{j-1}/2$. Thus $d(K_j(x_1), K_j(x_2)) > 2\sum_{i=j}^{\infty} \varepsilon_i$ and $d(H_j(x_1), H_j(x_2)) > 2\sum_{i=j}^{\infty} \tilde{\varepsilon}_i$ and by the Lemma for Proposition II.51, we get that $\lim_{i\to\infty} H_i = H$ and $\lim_{i\to\infty} K_i = K$ are each homomorphisms carrying M^n onto itself. Clearly $d(H$, identity$)$ and $d(K$, identity$) < \sum_{i=1}(\tilde{\varepsilon}/2^{i+1}) = \tilde{\varepsilon}/2 \leqslant \varepsilon/2$. We also have constructed the P_i's and $\tilde{P}_i$'s so that $\lim_{i\to\infty} P_i = \mathcal{P} = \lim_{i\to\infty} \tilde{P}_i$. Hence $H(\tilde{P}) = \mathcal{P} = K(P)$ and $h = K^{-1} \circ H$ is a homomorphism carrying M^n onto itself such that $h|\tilde{P} = \varphi|\tilde{P}$ and hence $h|\tilde{P}: \tilde{P} \longrightarrow M^n$ is a pwl homomorphism. Certainly $h|\tilde{L}$ = identity and the fact that we have obtained our desired ε-push h of $(M^n, \tilde{P}^k - \tilde{L})$ follows from the epsilontics involved.

To get that we have an isotopy connecting the identity and h we recall that for every i we had an isotopy $\tilde{h}_{t,i}$ $(0 \leqslant t \leqslant 1)$ connecting the identity and $\tilde{h}_i$. Let $\tilde{H}_t$ $(0 \leqslant t < \infty)$ be defined as $\tilde{H}_t(x) = \tilde{h}_{t-(i-1),i} \circ H_{i-1}(x)$ for $i-1 \leqslant t \leqslant i$. We note that for all positive integers i, that $\tilde{H}_i(x) = H_i(x)$. Similarly we obtain $\tilde{K}_t (0 \leqslant t < \infty)$ so that $\tilde{K}_i = K_i$ for each integer i. Then the desired isotopy connecting the identity and h is defined as follows:

$$h_t = \begin{cases} \tilde{H}_{t/\frac{1}{2}-t} & 0 \leqslant t < \frac{1}{2} \\ H & t = \frac{1}{2} \\ \tilde{K}^{-1}_{t-\frac{1}{2}/1-t} \circ H & \frac{1}{2} \leqslant t < 1 \\ K^{-1} \circ H & t = 1 \end{cases}$$

Proof of Theorem II.46: We have $|\tilde{M}^n| \subset |M^n|$ and since $\mathrm{CL}(\tilde{P}^k - \tilde{L})$ is compact we can cover $\mathrm{CL}(\tilde{P}^k - \tilde{L})$ with a finite number of open sets $\tilde{E}^n_1, \ldots, \tilde{E}^n_m$ so that $\mathrm{Cl}\, \tilde{E}^n_i \subset \tilde{M}^n$ for each i and so that there are subpolyhedra $\tilde{Q}^k_1, \ldots, \tilde{Q}^k_m$ in $\mathrm{Cl}(\tilde{P}^k - \tilde{L})$ with $\cup^m_{i=1} \tilde{Q}^k_i = \mathrm{Cl}(\tilde{P}^k - \tilde{L})$ and for each i, $\tilde{Q}^k_i \subset \tilde{E}^n_i$. Given $\varepsilon > 0$, let $\varepsilon' < \min(\varepsilon/_m, 1/m\, d(\cup^m_{i=1} \mathrm{Cl}\, \tilde{E}^n_i, M^n - \tilde{M}^n))$. We now consider the subpolyhedron $(\tilde{L} \cap \tilde{E}^n_1) \cup \tilde{Q}^k_1$ and apply Proposition II.51 using $\tilde{E}^n_1$ and this subpolyhedron. Also the epsilon we use there is $\varepsilon_1 < \min(\varepsilon', \tfrac{1}{2}\, d(\tilde{Q}^k_1, M^n - \tilde{E}^n_1))$. Hence we obtain an $h_1\colon M^n \longrightarrow M^n$ such that h_i = identity on $(\tilde{L} \cap \tilde{E}^n_1) \cup (M^n - U_{\varepsilon_1}(\tilde{Q}^k_1))$, $h_1 | \tilde{L} \cup \tilde{Q}^k_1 : \tilde{L} \cup \tilde{Q}^k_1 \longrightarrow M^n$ is pwl and h_1 is isotopic to the identity.

The next polyhedron we consider is $\{[h_1(\tilde{L} \cup \tilde{Q}^k_1)] \cap h_1(\tilde{E}^n_2)\} \cup h_1(\tilde{Q}^k_2) \subset h_1(\tilde{E}^n_2)$. The combinatorial triangulation we now take on $h_1(\tilde{E}^n_2)$ is that induced by h_1 on $h_1(\tilde{M}^n) \subset M^n$. Also we now have that $h_1(\tilde{L} \cup \tilde{Q}^k_1)$ is pwl in both M^n and $h_1(\tilde{M}^n)$ and hence we can apply Proposition II.51 again. Let $\varepsilon_2 < \min(\varepsilon', \tfrac{1}{2}\, d(h_1(\tilde{Q}^k_2), M^n - h_1(\tilde{E}^n_2)))$ and we obtain an $h_2\colon M^n \longrightarrow M^n$ such that h_2 = identity on $h_1(\tilde{L} \cup \tilde{Q}^k_1) \cup (M^n - U_{\varepsilon_2}(h_1(\tilde{Q}^k_2)))$, $h_2 |_{\tilde{L} \cup h_1(\tilde{Q}^k_1) \cup h_1(\tilde{Q}^k_2)} : \tilde{L} \cup h_1(\tilde{Q}^k_1) \cup h_1(\tilde{Q}^k_2) \longrightarrow M^n$ is pwl and h_2 is isotopic to the identity.

Hence by merely repeating this process m-times the result follows. Also we have picked our ε_i's so that $\varepsilon_1 + \varepsilon_2 + \ldots + \varepsilon_m < \varepsilon$.

Definitions. Let X denote a metric space and $A \subset X$ a subset whose closure $\bar{A}$ is compact. Let M denote a topological manifold with a complete metric d. Hom $(X, A; M)$ will denote an arbitrary set of embeddings of X into M, all of which agree on $X - A$. That is, if $f, g \in \mathrm{Hom}(X, A; M)$, then $f | X - A = g | X - A$. If $f, g \in \mathrm{Hom}(X, A; M)$, we define a distance function $d(f, g) = \text{l.u.b.}_{x \in X}\, d(f(x), g(x)) =$

$\text{l.u.b.}_{x \in \bar{A}}\, d(f(x), g(x))$, which exists because $\bar{A}$ is compact. This distance function makes $\text{Hom}(X, A; M)$ into a metric space.

Let F be a subset of $\text{Hom}(X, A; M)$ with the property that for each $g \in \text{Hom}(X, A; M)$ and each real number $\varepsilon > 0$ there is an $f \in F$ with $d(f, g) < \varepsilon$. We will say F is dense in $\text{Hom}(X, A; M)$.

Now let F be a subset of $\text{Hom}(X, A; M)$ with the following property: for any $\varepsilon > 0$ there is a $\delta > 0$ such that if $f, f' \in F$ and $d(f, f') < \delta$, then there is an ε push h of $(M, f(A))$ such that $h \circ f = f'$. Then we say that F is solvable.

LEMMA II.52. *Let $F \subset F' \subset \text{Hom}(X, A; M)$. Suppose that for each $f' \in F'$ and each $\varepsilon > 0$ there is an ε-push h of $(M, f'(A))$ such that $f = h \circ f'$ is an element of F. If F is solvable, then F' is also solvable.*

Proof. Since F is solvable, given $\varepsilon/3$ there is a δ_1 so that if $f_1, f_2 \in F$ with $d(f_1, f_2) < \delta_1$ then were is an $\varepsilon/3$-push h of $(M, f_1(A))$ such that $h \circ f_1 = f_2$. Let the δ for the solvability of F' be $\delta_1/3$ and suppose that $f_1', f_2' \in F'$ and $d(f_1', f_2') < \delta = \delta_1/3$. Let $\varepsilon_1 < \min(\delta_1/3, \varepsilon/3)$. By hypotheses for $f_i' \in F'$ $(i = 1, 2)$ and $\varepsilon_1 > 0$ there is an ε_1-push h_i of $(M, f_i'(A))$ such that $f_i = h_i \circ f_i'$ is an element of F. Since $d(f_1', f_2') < \delta_1/3$ and h_i $(i = 1, 2)$ is an ε_1-push with $\varepsilon_1 < \delta_1/3$ it follows that $d(f_1, f_2) < \delta_1$. Thus there is an $\varepsilon/3$-push h of $(M, f_1(A))$ such that $h \circ f_1 = f_2$. The desired ε-push of $(M, f_1'(A))$ carrying f_1' to f_2' is then $h_2^{-1} \circ h \circ h_1$.

The main proposition we will want in obtaining Theorems II.47- II.49 is as follows.

PROPOSITION II.53. *The union of two dense, solvable subsets of $\text{Hom}(X, A; M)$ is dense and solvable.*

Proof. The fact that the union is dense is trivial. Suppose F, F' are the two dense solvable subsets. It will suffice to show that if given $\varepsilon > 0$ then there exists a $\delta > 0$ so that if $d(f_1, f_1') < \delta$, where $f_1 \in F$ and $f_1' \in F'$, then there is an ε-push h of $(M, f_1(A))$ such that $hf_1 = f_1'$. The proof is similar to that used in proving Proposition II.51. That is we have to obtain two sequences of homomorphisms $\{K_i\}$ and $\{H_i\}$ so as to be able to apply the lemma for Proposition II.51 and so that $K^{-1} \circ H = h$ gives the ε-push of $(M, f_1(A))$ with $hf_1 = f_1'$. The fact that each of F and F' is dense and solvable allows us to obtain the desired sequences of homomorphisms $\{H_i\}$ and $\{K_i\}$. We now give some details.

Let $\varepsilon > 0$ be given and let $\varepsilon_1' = \min(\varepsilon/4, \frac{1}{8})$. We take δ to be $\delta_1'/2$ where δ_1' is a positive number so that if $f_1', f_2' \in F'$ and $d(f_1', f_2') < \delta_1'$, then there is an ε_1'-push h_1' with $f_2' = h_1' \circ f_1'$. We will denote this δ_1' by $\delta_1' = \delta(F', \varepsilon_1')$. Let $\varepsilon_1 = \min(\varepsilon/4, \frac{1}{8})$ and $\delta_1 = \delta(F, \varepsilon_1)$. Thus if $f_1, f_2 \in F$ and $d(f_1, f_2) < \delta_1$ then there is an ε_1-push h_1 with $f_2 = h_1 \circ f_1$.

Now suppose $f_1' \in F$ and $f_1 \in F$ and $d(f_1', f_1) < \delta = \delta_1'/2$. Since F' is dense, there is an $f_2' \in F'$ so that $d(f_1, f_2') < \min(\delta_1'/2, \delta_1/2)$. Since $d(f_1, f_1') < \delta_1'/2$ and $d(f_1, f_2') < \delta_1'/2$, we have that $d(f_1', f_2') < \delta'$ and hence there is an ε_1'-push h_1' with $f_2' = h_1' \circ f_1'$. Let $\eta_1' = \text{g.l.b.}\{d(h_1'(x), h_1'(y)) \mid d(x, y) \geqslant \frac{1}{2}\}$ and let $\varepsilon_2' = \min(\varepsilon_1'/2, \eta_1'/8)$ and $\delta_2' = \delta(F, \varepsilon_2')$. Since F is dense there is an $f_2 \in F$ so that $d(f_2', f_2) < \min(\delta_1/2, \delta_1'/2)$. Since $d(f_1, f_2') < \delta_1/2$ and $d(f_1, f_2) < \delta_1$, it follows that there is an ε_1-push h_1 with $f_2 = h_1 \circ f_1$. Let $\eta_1 = \text{g.l.b.}\{d(h_1(x), h_1(y)) \mid d(x, y) \geqslant \frac{1}{2}\}$.

We now want to define f_i', f_i, h_{i-1}' and h_{i-1} inductively. Suppose for some $i \geqslant 3$ we have f_i' and f_i defined for $1 \leqslant j \leqslant i-1$ and h_{j-1}' and h_{j-1} defined for $2 \leqslant j \leqslant i-1$ so that for $2 \leqslant j \leqslant i-1$ we have the following conditions satisfied:

(1) $d(f_j', f_{j-1}) < \min(\delta_{j-1}'/2, \delta_{j-1}/2)$;

(2) $d(f_j, f_j') < \min(\delta_{j-1}/2, \delta_j'/2)$;

(3) there is an ε_{j-1}'-push h_{j-1}' with $f_j' = h_{j-1}' \circ f_{j-1}'$;

(4) there is an ε_{j-1}-push h_{j-1} with $f_j = h_{j-1} \circ f_{j-1}$;

(5) $\eta_{j-2}' = \text{g.l.b.}\,\{d(h_{j-2}' \circ h_{j-3}' \circ \ldots \circ h_1'(x), h_{j-2}' \circ \ldots \circ h_1'(y)) \mid d(x,y) \geqslant 1/_{j-1}\}$, $\varepsilon_{j-1}' = \min(\varepsilon_{j-2}'/2, \eta_{j-2}'/8)$ and $\delta_{j-1}' = \delta(F', \varepsilon_{j-1}')$ (we define $\varepsilon_0' = \varepsilon/2$ and $\eta_0' = 1$);

(6) $\eta_{j-2} = \text{g.l.b.}\,\{d(h_{j-2} \circ \ldots \circ h_1(x), h_{j-2} \circ \ldots \circ h_1(y)) \mid d(x,y) \geqslant 1/_{j-1}\}$, $\varepsilon_{j-1} = \min(\varepsilon_{j-2}/2, \eta_{j-2}/8)$ and $\delta_{j-1} = \delta(F, \varepsilon_{j-1})$ (we define $\varepsilon_0 = \varepsilon/2$ and $\eta_0 = 1$); and

(7) $\delta_j' = \delta(F', \varepsilon_j')$ where $\varepsilon_j' = \min(\varepsilon_{j-1}'/2, \eta_{j-1}'/8)$ and $\eta_{j-1}' = \text{g.l.b.}\,\{d(h_{j-1}' \circ h_{j-2}' \circ \ldots \circ h_1'(x), h_{j-1}' \circ h_{j-2}' \circ \ldots \circ h_1'(y)) \mid d(x,y) \geqslant 1/j\}$.

The following diagram illustrates the above facts where the dotted arrows point to the new approximating embeddings; the vertical arrows indicate the pushes; and the numbers above or alongside the arrows give an upper bound on the distance between the images of points under the two embeddings connected by the arrows.

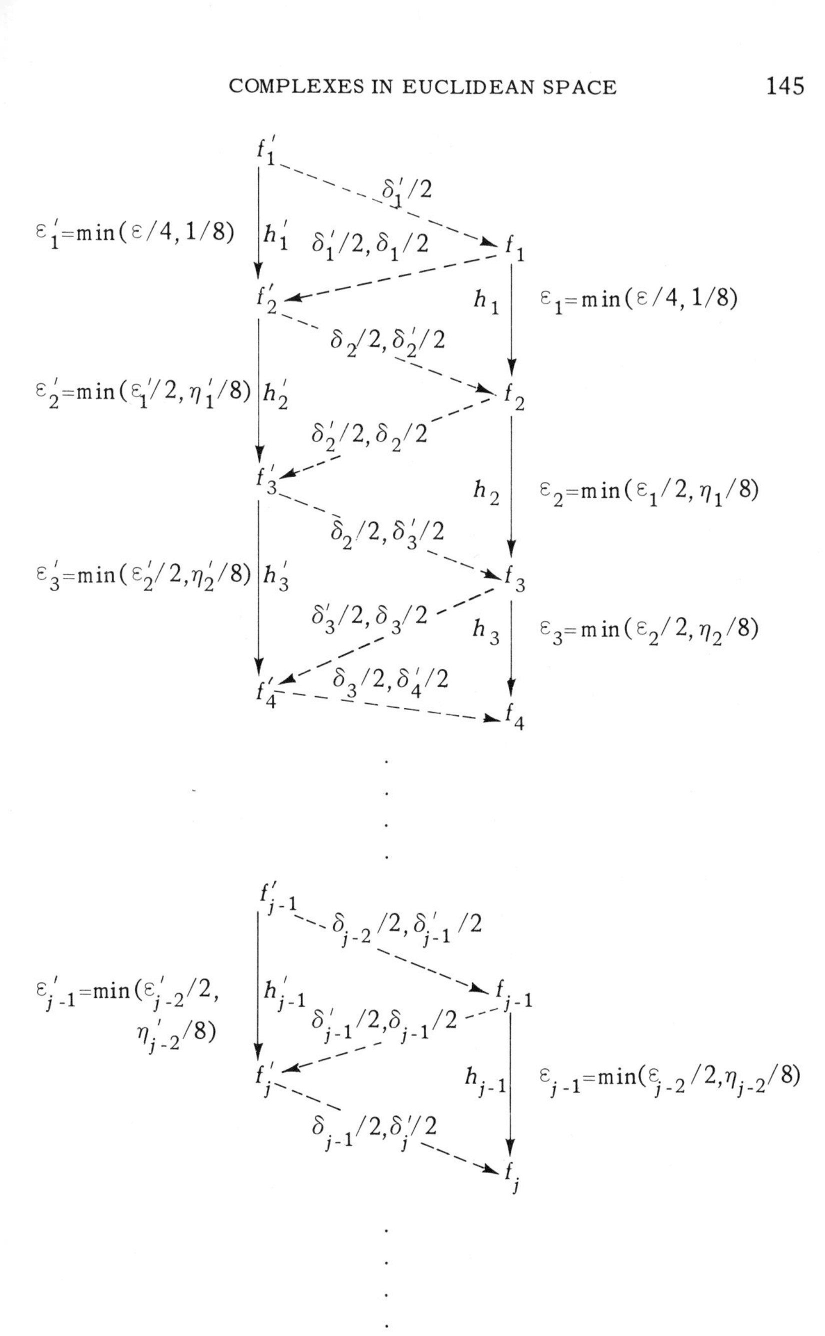
f_1'
$\delta_1'/2$
$\varepsilon_1' = \min(\varepsilon/4, 1/8)$
h_1'
$\delta_1'/2, \delta_1/2$
f_1
f_2'
h_1
$\varepsilon_1 = \min(\varepsilon/4, 1/8)$
$\delta_2/2, \delta_2'/2$
$\varepsilon_2' = \min(\varepsilon_1'/2, \eta_1'/8)$
h_2'
f_2
$\delta_2'/2, \delta_2/2$
f_3'
h_2
$\varepsilon_2 = \min(\varepsilon_1/2, \eta_1/8)$
$\delta_2/2, \delta_3'/2$
$\varepsilon_3' = \min(\varepsilon_2'/2, \eta_2'/8)$
h_3'
f_3
$\delta_3'/2, \delta_3/2$
h_3
$\varepsilon_3 = \min(\varepsilon_2/2, \eta_2/8)$
$\delta_3/2, \delta_4'/2$
f_4'
f_4
f_{j-1}'
$\delta_{j-2}/2, \delta_{j-1}'/2$
$\varepsilon_{j-1}' = \min(\varepsilon_{j-2}'/2,$
$\eta_{j-2}'/8)$
h_{j-1}'
f_{j-1}
$\delta_{j-1}'/2, \delta_{j-1}/2$
f_j'
h_{j-1}
$\varepsilon_{j-1} = \min(\varepsilon_{j-2}/2, \eta_{j-2}/8)$
$\delta_{j-1}/2, \delta_j'/2$
f_j

We now define f_i', f_i, h_{i-1}' and h_{i-1}. As above $\eta_{i-2}' =$ g.l.b. $\{d(h_{i-2}' \circ \ldots \circ h_1'(x), h_{i-2}' \circ \ldots \circ h_1'(y)) \mid d(x, y) \geqslant 1/_{i-1}\}$, $\varepsilon_{i-1}' = \min(\varepsilon_{i-2}'/2, \eta_{i-2}'/8)$ and $\delta_{i-1}' = \delta(F', \varepsilon_{i-1}')$. Let $\eta_{i-2} =$ g.l.b. $\{d(h_{i-2} \circ \ldots \circ h_1(x), h_{i-2} \circ \ldots \circ h_1(y)) \mid d(x, y) \geqslant 1/_{i-1}\}$, $\varepsilon_{i-1} = \min(\varepsilon_{i-2}/2, \eta_{i-2}/8)$ and $\delta_{i-1} = \delta(F, \varepsilon_{i-1})$. Let $f_i' \in F'$ approximate f_{i-1} so that $d(f_i', f_{i-1}) < \min(\delta_{i-1}'/2, \delta_{i-1}/2)$. Then we have that $d(f_i', f_{i-1}') < \delta_{i-1}'$ and hence there is an ε_{i-1}'-push h_{i-1}' with $f_i' = h_{i-1}' \circ f_{i-1}'$. Let $\eta_{i-1}' =$ g.l.b. $\{d(h_{i-1}' \circ \ldots \circ h_1'(x), h_{i-1}' \circ \ldots \circ h_1'(y)) \mid d(x, y) \geqslant 1/i\}$, $\varepsilon_i' = \min(\varepsilon_{i-1}'/2, \eta_{i-1}'/8)$ and $\delta_i' = \delta(F', \varepsilon_i')$. Let $f_i \in F$ approximate f_i' so that $d(f_i, f_i')$. $< \min(\delta_{i-1}/2, \delta_i'/2)$. Then we have that $d(f_{i-1}, f_i) < \delta_{i-1}$ and hence there is an ε_{i-1}-push h_{i-1} with $f_i = h_{i-1} \circ f_{i-1}$. This then completes our inductive step.

For each $i \geqslant 1$ let $K_i = h_i' \circ h_{i-1}' \circ \ldots \circ h_1'$ and $H_i = h_i \circ h_{i-1} \circ \ldots \circ h_1$. The claim is that the two sequences of homomorphisms $\{K_i\}$ and $\{H_i\}$ now allow us to obtain an ε-push h so that $h \circ f_1 = f_1'$. Clearly $d(K_i, K_{i+1}) < \varepsilon_i'$ since $d(\text{identity}, h_{i+1}') < \varepsilon_{i+1}' \leqslant \varepsilon_i'/2$ and $d(H_i, H_{i+1}) < \varepsilon_i$ since $d(\text{identity}, h_{i+1}) < \varepsilon_{i+1} \leqslant \varepsilon_i/2$. Let $K = \lim i \to \infty\, K_i$ and $H = \lim i \to \infty\, H_i$. Since $d(f_i', f_i) \to 0$ as $i \to \infty$, $h_i' \circ f_i' = f_{i+1}'$ and $h_i \circ f_i = f_{i+1}$, it follows that $K \circ f_1' = H \circ f_1$. If $j \geqslant 2$ and $d(x_1, x_2) > 1/j$ then $d(K_{j-1}(x_1), K_{j-1}(x_2)) > \eta_{j-1}'$ and $d(H_{j-1}(x_1), H_{j-1}(x_2)) > \eta_{j-1}$. Since $d(h_j', \text{identity}) < \eta_{j-1}'/8$ and $d(h_j, \text{identity}) < \eta_{j-1}/8$, it follows that $d(K_j(x_1), K_j(x_2)) > \eta_{j-1}' - \eta_{j-1}'/8 - \eta_{j-1}'/8 = 3\eta_{j-1}'/4$ and that $d(H_j(x_1), H_j(x_2)) > 3\eta_{j-1}/4$.

Now $2\Sigma_{i=j}^{\infty}\, \varepsilon_i' \leqslant 2\Sigma_{i=1}^{\infty}\, \eta_{j-1}'/8 \cdot 1/2^{i-1} = \eta_{j-1}'/2$ and similarly $2\Sigma_{i=j}^{\infty}\, \varepsilon_i \leqslant \eta_{j-1}/2$. Thus $d(K_j(x_1), K_j(x_2)) > 2\Sigma_{i=j}^{\infty}\, \varepsilon_i'$ and $d(H_j(x_1), H_j(x_2)) > 2\Sigma_{i=j}^{\infty}\, \varepsilon_i$ and K and H are homomorphisms. Also $d(\text{identity}, K) < \Sigma_{i=1}^{\infty}\, \varepsilon_i' < \Sigma_{i=2}^{\infty}\, \varepsilon/2^i = \varepsilon/2$ and $d(\text{identity}, H) < \Sigma_{i\ 1}^{\infty}\, \varepsilon_i < \Sigma_{i\ 2}^{\infty}\, \varepsilon/2^i = \varepsilon/2$. Setting $h = K^{-1} \circ H$ we have that

$d(\text{identity}, h) < \varepsilon$ and $hf_1 = K^{-1} \circ Hf_1 = K^{-1} \circ Kf_1' = f_1'$. Thus h is an ε-push carrying f_1 to f_1'.

We now return to the proof of Theorem II.47. We first prove this theorem with the following temporary restriction: the triangulation of P^k must be one in terms of which f is locally tame (this restriction will be removed shortly).

Referring back to the definition of 'locally tame', call a subset $A \subset P^k$ small if there is a neighborhood U of A in P^k and a neighborhood V of $f(A)$ in M^n such that (1) $f(U) = V \cap f(P^k)$ and (2) V has a triangulation as a combinatorial manifold in terms of which $f|U : U \longrightarrow V$ is pwl with respect to an induced triangulation of U as an open subset of P^k.

Observe that a finite disjoint union of compact small subsets of P^k is itself small, the neighborhoods U and V for the union simply being disjoint unions of sufficiently small corresponding neighborhoods for the individual sets.

Taking a preliminary subdivision if necessary, assume P^k to be triangulated so finely that the stars are simplexes are small. Let N_i denote the closed regular neighborhood of $P^{(i)}$, the i-skeleton of P^k, in the second barycentric subdivision of P^k. Then $N_0 \subset N_1 \subset \ldots \subset N_k = P^k$ and by the above observation we easily obtain that N_0 is small and that $\mathrm{Cl}(N_i - N_{i-1})$, the closure of $N_i - N_{i-1}$ is small for $i = 1, \ldots, k$. The plan is to define, with the help of Theorem II.46, $k+1$ homomorphisms of $M^n, h_0, h_1, \ldots, h_k$, with the properties:

(i) h_0 is an $(\varepsilon/_{k+1})$-push of $(M^n, f(N_0))$ such that $h_0 f|N_0 : N_0 \longrightarrow M^n$ is pwl.

(ii) h_0 is an $(\varepsilon/_{k+1})$-push of $(M^n, h_{i-1} h_{i-2} \circ \ldots \circ h_0 \circ f(\mathrm{Cl}(N_i - N_{i-1})))$ such that $h_i h_{i-1} \ldots h_0 f|N_i : N_i \longrightarrow M^n$ is pwl, for $i = 1, 2, \ldots, k$.

Setting $h = h_k h_{k-1} \cdots h_0$ will then complete the proof.

Step 0. Since N_0 is small, there is a neighborhood U_0 of N_0 in P^k and a neighborhood V_0 of $f(N_0)$ in M^n satisfying (1) and (2) above in the definition of small subset. To apply Theorem II.46, let

$M^n = M^n$ with its given preassigned combinatorial triangulation,

$\tilde{M}^n = V_0$, triangulated so that $f|U_0 : U_0 \longrightarrow V_0$ is pwl,

$\tilde{P}^k = f(N_0)$, with the triangulation carried over from N_0 by f,

$\tilde{L} = \emptyset$, the empty set, and

$\varepsilon \leqslant$ the present $\varepsilon/k+1$.

Theorem II.46, then asserts the existence of an $(\varepsilon/k+1)$-push h_0 of $(M^n, f(N_0))$ such that $h_0 f|N_0 : N_0 \longrightarrow M_0^n$ is pwl. This completes Step 0.

Step 1. Since $\mathrm{Cl}(N_1 - N_0)$ is small, there is a neighborhood U_1 of $\mathrm{Cl}(N_1 - N_0)$ in P^k and a neighborhood V_1 of $f(\mathrm{Cl}(N_1 - N_0))$ in M^n satisfying conditions (1) and (2) above. Now to apply Theorem II.46, let

$M^n = M^n$ with its given preassigned combinatorial triangulation,

$\tilde{M}^n = h_0(V_1)$, triangulated so that $h_0 f|U_1 : U_1 \longrightarrow h_0(V_1)$ is pwl,

$\tilde{P}^k = h_0 f(N_1 \cap U_1)$, with the triangulation carried over from $N_1 \cap U_1$ by $h_0 f$,

$\tilde{L} = h_0 f(N_0 \cap U_1)$, and

$\varepsilon =$ the present $\varepsilon/k+1$.

Theorem II.46 then asserts the existence of an $(\varepsilon/k+1)$-push h_1 of $(M^n, h_0 f(\mathrm{Cl}(N_1 - N_0))$ such that $h_1 h_0 f|N_1 : N_1 \longrightarrow M^n$ is pwl, provided that we also restrict ε so small that h_1 restricts to the identity outside $h_0(V_1)$. This completes Step 1.

Steps 2 through k are then performed in a similar manner, and the proof is completed by setting $h = h_k h_{k-1} \cdots h_0$.

Proof of Theorem II.47. Let $\mathrm{Hom}(P^k, M^n)$ denote the set of all locally tame embeddings of P^k into M^n, and let $2k+2 \leqslant n$. Let $(P^k)_1$ and $(P^k)_2$ denote P^k with two different triangulations, and let M^n have a fixed but arbitrary combinatorial triangulation. Let F_1 denote the set of pwl embeddings of $(P^k)_1$ into M^n, and F_1' the set of embeddings of P^k into M^n which are locally tame with respect to the triangulation $(P^k)_1$. Similarly for F_2 and F_2' with respect to $(P^k)_2$. By general position, it follows that each of F_1 and F_2 are dense. By using Proposition II.50 locally as we did in obtaining Theorem II.46 from Proposition II.51, it follows that each of F_1 and F_2 are solvable. Now by Lemma II.52 and the above 'restricted' version of Theorem II.47, it follows that each of F_1' and F_2' are also dense and solvable.

Therefore, by Proposition II.53, $F_1' \cup F_2$ is dense and solvable. Let $f_1' \in F_1'$. Since F_2 is dense in $\mathrm{Hom}(P^k, M^n)$, choose $f_2 \in F_2$ so that $d(f_1', f_2) < \delta_{F_1' \cup F}(\epsilon)$. Then there is an ε-push h of $(M^n, f_1'(P^k))$ such that $hf_1' = f_2$. Since f_1' was locally tame with respect to $(P^k)_1$, while f_2 is pwl with respect to $(P^k)_2$, this completes the proof of Theorem II.47.

As a corollary to Theorem II.47, we get the following result.

THEOREM II.54. *Let f be an embedding of the polyhedron P^k into the combinatorial manifold M^n. If $2k+2 \leqslant n$ and if f is locally tame with respect to one triangulation of P^k, then f is locally tame with respect to any triangulation of P^k.*

Proof of Theorem II.54. Suppose $f\colon P^k \longrightarrow M^n$ is a locally tame embedding with respect to $(P^k)_1$ say. Consider an arbitrary triangulation of P^k, say $(P^k)_2$. Then by Theorem II.47 there exists an ε-push h such that $hf\colon P^k \longrightarrow M^n$ is pwl with respect to $(P^k)_2$. Then $h^{-1}(M^n)$ gives a triangulation containing $f(P^k)$ so that $f\colon P^k \longrightarrow M^n$ is locally tame with respect to $(P^k)_2$.

Proof of Theorem II.48. Suppose $f\colon P^k \longrightarrow M^n$ is locally tame with respect to $(P^k)_1$ and $f'\colon P^k \longrightarrow M^n$ is locally tame with respect to $(P^k)_2$. Taking $\mathrm{Hom}(P^k, M^n)$, F_1' and F_2' as in the proof of Theorem II.47, it follows by Theorem II.54 that $F_1' = \mathrm{Hom}(P^k, M^n) = F_2'$. Since this is solvable, Theorem II.48 follows.

Proof of Theorem II.49. Now, since $\mathrm{Hom}(P^k, M^n)$ is solvable when $2k+2 \leqslant n$, let $\delta = \delta(1)$ be the delta determined by the solvability for $\varepsilon = 1$. Suppose that f, $f' \in \mathrm{Hom}(P^k, M^n)$ are homotopic as maps of P^k into M^n. Then there is a sequence of continuous maps $f = g_0, g_1, \dots, g_r = f'$ of P^k into M such that $d(g_i, g_{i+1}) < \delta$. $\mathrm{Hom}(P^k, M^n)$, since it includes the pwl embeddings, is dense in the mapping space $(M^n)^{P^k}$, and hence there is also a sequence of elements of $\mathrm{Hom}(P^k, M^n)$, $f = f_0, f_1, \dots, f_r = f'$, such that $d(f_i, f_{i+1}) < \delta$. According to the definition of δ, there exist 1-pushes h_i such that $h_i f_{i-1} = f_i$ for $i = 1, \dots, r$. Then $h = h_r h_{r-1} \cdots h_1$ is a homomorphism of M^n onto itself which is isotopic to the identity, such that $hf = f'$, and Theorem II.49 follows.

We now give two applications of the theory of this section. An embedding f of a topological manifold M^k into a topological manifold M^n is said to be locally flat if for each $x \in M^k$ there is a neighborhood U of $f(x)$ in M^n such that the pair $(U, U \cap f(M^k))$ is homomorphic to the pair (E^n, E^k).

THEOREM II.55. *Let f be a locally flat embedding of the closed combinatorial manifold M^k into the combinatorial manifold M^n. If $2k+2 \leqslant n$, then for each $\varepsilon > 0$ there is an ε-push of $(M^n, f(M^k))$ such that $hf \colon M^k \longrightarrow M^n$ is pwl.*

Proof. For each $x \in M^k$, choose a neighborhood U of $f(x)$ in M^n as in the above definition. $U \cap f(M^k)$ inherits a triangulation as an open combinatorial manifold from M^k via f. Since $(U, U \cap f(M^k))$ is homomorphic to (E^n, E^k), this triangulation extends to a 'product' triangulation of U as a combinatorial manifold, in terms of which $f|_{f^{-1}(U \cap f(M^k))}$ is pwl. But then f is locally tame, and an application of Theorem II.47 completes the proof.

THEOREM II.56. *Let f and f′ be locally tame embeddings of the polyhedron P^k into the n-sphere S^n (or Euclidean space E^n). If $2k+2 \leqslant n$, there is a homomorphism h of S^n (or E^n) onto itself which is isotopic to the identity, such that $hf = f'$.*

Proof. f must be homotopic to f' and them Theorem II.49 applies.

APPENDIX A

Locally Flat Embeddings of Topological Manifolds

The results of this section are due to Morton Brown, and appear in his paper 'Locally Flat Embeddings of Topological Manifolds', *Annals of Math* 75 (1962), 331-341.

Definitions. Let X be a topological space and $B \subset X$, then B is (bi-) collared in X if there exists a homomorphism $h: B \times [0,1)$ $(B \times (-2,1)) \longrightarrow V$ open $\subset X$ such that $B \subset V$ and $h(b,0) = b$ for every $b \in B$. If B can be covered by a collection of open sets (relative to B) each of which is (bi-) collared in X, then B is locally (bi-) collared in X.

The main result we want to obtain is as follows:

THEOREM A.1. *A locally collared subset of a separable metric space is collared.*

To prove this result it suffices to prove the following two lemmas:

LEMMA A.2. *Let B be a subset of a metric space X. Suppose $B = U_1 \cup U_2$ where U_1 and U_2 are open in B and $U_1 \cap U_2 \neq \emptyset$, then if both U_1 and U_2 are collared in X, then B is also.*

LEMMA A.3. *Suppose X is a separable metric space and $\mathcal{C}$ is a topological property such that the following three conditions are also true:*

(i) *if U open $\subset X$ has $\mathcal{C}$, then any open subset $V \subset U$ has $\mathcal{C}$,*

(ii) *if $\{U_\alpha\}$ $\alpha \in A$, is a pairwise disjoint collection of open subsets of X each having $\mathcal{C}$, then $\bigcup_{\alpha \in A} U_\alpha$ has $\mathcal{C}$, and*

(iii) *if U_1 and U_2 are open subsets of X, each having $\mathcal{C}$, then $U_1 \cup U_2$ has $\mathcal{C}$.*

Then if there exists an open cover $\{U_\alpha\}$ of X, with each U_α having $\mathcal{C}$, then X has $\mathcal{C}$.

We will first prove Lemma A.3, then assuming Lemma A.2 we will show why these imply Theorem A.1, and then obtain some important corollaries as applications to manifolds. The proof of Lemma A.2 involves proving four other lemmas first which are not of interest in themselves and since the lack of a proof of Lemma A.2 does not hinder us in our understanding of the other results, we omit the proof here and refer the reader to the above reference.

Proof of Lemma A.3. Since X is a separable metric space and $\{U_\alpha\}$ is an open cover of X there exists a countable subcover $\{0_i\}$ of X so that each 0_i has $\mathcal{C}$. Let $S^n = \bigcup_{i=1}^n 0_i$, by (iii) S^n has $\mathcal{C}$. Let $V_n = \{x \in X \mid d(x, X - S^n) > 1/n$, then $V_n \subset S^n$ and by (i) has $\mathcal{C}$ (also $\overline{V}_n \subset V_{n+1}$). Let $W_1 = V_1$, $W_2 = V_2$, $W_3 = V_3$ and $W_n = V_n - V_{n-3}$ ($n \geqslant 4$), then $W_n \subset V_n \subset S^n$ and hence each W_n also has $\mathcal{C}$ by (i). We now consider the two collections of open sets $\{W_{4n-1}\}_{n=1}^\infty$ and $\{W_{4n+1}\}_{n=0}^\infty$. Each are pairwise disjoint collections and by (ii), $\bigcup_{n=1}^\infty W_{4n-1}$ and $\bigcup_{n=0}^\infty W_{4n+1}$ both have $\mathcal{C}$. But now $X = (\bigcup_{n=1}^\infty W_{4n-1}) \cup (\bigcup_{n=0}^\infty W_{4n+1})$ and hence by (iii), X has $\mathcal{C}$.

Proof of Theorem A.1. Let B be a locally collared subset of X, then there exists an open $\{U_\alpha\}$ of B such that each U_α is collared in X. The $\mathcal{C}$ of Lemma A.3 will be the property of being collared in X. Since X is a separable metric space, B is also and hence we only need to verify conditions (i)-(iii) of Lemma A.3.

(i) is trivial, for any collar of U, restricted to V gives a collar of V. (iii) follows from Lemma A.2, if $U_1 \cap U_2 = \emptyset$ and from (ii) if $U_1 \cap U_2 \neq \emptyset$. (ii) is obtained as follows. Suppose h_α: $U_\alpha \times [0,1) \longrightarrow$ neighborhood of U_α in X such that $h_\alpha(x, 0) = x \in U_\alpha$. Since X is a metric space there exists a pairwise disjoint collection $\{W_\alpha\}_{\alpha \in A}$ of open subsets of X such that $U_\alpha \subset W_\alpha \subset h_\alpha(U_\alpha \times [0,1))$, $\alpha \in A$. For example let $W_\alpha = h_\alpha(U_\alpha \times [0,1)) \cap \{x \in X \mid d(x, U_\alpha) < d(x, \cup_{\beta \neq \alpha} U_\beta)\}$. Let $\mathcal{O} = \cup_{\alpha \in A} h_\alpha^{-1}(W_\alpha)$ then $\mathcal{O}$ is an open subset of $B \times [0,1)$ containing $\cup_{\alpha \in A} U_\alpha \times \{0\}$. For each α there exists t_α, $0 < t\alpha < 1$ such that $U_\alpha \times [0, t_\alpha) \subset \mathcal{O}$. Let k_α: $U_\alpha \times [0,1) \longrightarrow X$ be defined as $k_\alpha(u, t) = h_\alpha(u, tt_\alpha)$ and the desired collar of $\cup_{\alpha \in A} U_\alpha$ is obtained as h: $\cup_{\alpha \in A} U_\alpha \times [0,1) \longrightarrow X$ where $h \mid U_\alpha \times [0,1) = k_\alpha$.

Of interest to us will be the following applications of Theorem A.1 to manifolds.

LEMMA A.4. *The boundary of an n-manifold with boundary is locally collared.*

This follows from the definition of n-manifold with boundary, i.e., each $p \in \mathring{M}$ has a closed neighborhood homomorphic to I^n.

Definition. A $(n-1)$-dimensional submanifold B of an n-manifold X is locally flat if for each $b \in B$ there is a neighborhood N_b of b in X and a homomorphism $h_b : N_b \longrightarrow E^n$ such that $h_b(N_b \cap B) \subset E^{n-1} \subset E^n$. The following lemma is easily established.

LEMMA A.5. *A submanifold B^{n-1} of a manifold X^n is locally flat in X^n if and only if it is bi-collared in X^n.*

THEOREM A.6. *The boundary of an n-manifold with boundary is collared.*

This follows directly from Theorem A.1 and Lemma A.4.

THEOREM A.7. *Let B^{n-1} be a locally flat two-sided $(n-1)$-submanifold of a manifold X^n. Then B^{n-1} is bi-collared in X^n.*

Proof. Let N be a connected neighborhood of B in X which is separated by B, and let Q, R be the components of $N-B$. Since B is locally flat in N, $Q \cup B$ and $R \cup B$ are manifolds with boundary B. It follows from Theorem A.6 that B is collared in each. Hence B is bi-collared in X.

THEOREM A.8. *Let Σ^{n-1} be locally flat in S , then Σ^{n-1} is flat in S (i.e., there exists a homomorphism h: $S^n \longrightarrow > S^n$ such that $h(\Sigma^{n-1}) = S^{n-1} \subset S^n$).*

Proof. This follows from Theorem A.7 and the generalized Schoenflies theorem.

Definitions: A 0-star sphere Σ^0 is a pair of points. A 0-star cell $\mathfrak{J}^0$ is a single point. For $n > 0$ an n-star sphere Σ^n (n-star cell $\mathfrak{J}^n$) is a finite complex homomorphic to the n-sphere S^n(n-cell I^n) and such that the link of each vertex is a Σ^{n-1}(Σ^{n-1} or $\mathfrak{J}^{n-1}$). An n-star manifold M^n (manifold N^n with boundary) is a locally finite complex such that the link of each vertex is a Σ^{n-1} (Σ^{n-1} or $\mathfrak{J}^{n-1}$). A 0-star manifold (manifold with boundary) is an even- (odd) numbered set of points. Any combinatorial n-manifold is an n-star manifold.

THEOREM A.9. *Let M^{n-1} be an $(m-1)$-star manifold embedded as a subcomplex of an n-star manifold M^n. Then M^{n-1} is locally flat in M^n.*

Proof. The theorem is evidently true for $n = 1$. Inductively, suppose we have proved the theorem for $n=k$. Let M^k be a k-star manifold embedded as a subcomplex of the $(k+1)$-star manifold M^{k+1}. Let v be a vertex of M^k. Then $\mathrm{lk}\,(v, M^k)$ is a Σ^{k-1} embedded as a subcomplex of $\mathrm{lk}(v, M^{k+1})$, which is a Σ^k. By the induction hypothesis, $\mathrm{lk}(v, M^k)$ is locally flat in $\mathrm{lk}(v, M^{k+1})$. Applying Theorem A.8 we obtain a homeomorphism $h\colon \mathrm{lk}(v, M^{k+1}) \longrightarrow S^k$ such that $h\,(\mathrm{lk}(v, M^k))$ is the equator S^{k-1} of S^k. We may think of S^k as the unit sphere of E^{k+1} with S^{k-1} in the hyperplane E^k. Since $\mathrm{st}(v, M^{k+1})$ is the join of v and $\mathrm{lk}(v, M^{k+1})$ and since the unit ball B^{k+1} is the join of the origin and S^k, h can be extended in the obvious way to a homomorphism $\bar{h}\colon \mathrm{st}(v, M^{k+1}) \longrightarrow B^{k+1}$. Furthermore, $\mathrm{st}(v, M^k)$ is the join of v with $\mathrm{lk}(v, M^k)$. Hence $\bar{h}\,[\mathrm{st}(v, M^k)] \subset E^k$. Since each point of M^k lies in the interior of the star of some vertex of M^k, we have established that M^k is locally flat in M^{k+1}.

THEOREM A.10. *Let M^{n-1} be an $(n-1)$-star manifold embedded as a two-sided subcomplex of an n-star manifold M^n. Then M^{n-1} is bi-collared in M^n.*

The result is an immediate consequence of Theorems A.9 and A.7.

THEOREM A.11. *Let Σ^{n-1} be an $(n-1)$-star sphere embedded as a subcomplex of an n-star triangulation of the n-sphere S^n. Then Σ^{n-1} is flat in S^n.*

This follows from Theorems A.9 and A.8.

APPENDIX B

Some Remarks on Knot Theory

The following discussion is intended to indicate briefly how one computes the fundamental group of the complement of a knot in 3-space (or in S^3). For further details the interested reader can refer to R.H. Fox, 'A quick trip through knot theory', *Topology of 3-manifolds*, Prentice-Hall (1962), pp. 120–167; or to R.H. Crowell and R.H. Fox, *Introduction to knot theory*, Ginn and Company (1963).

By a knot we will mean a homomorphism h taking $S^1 \longrightarrow E^3$. A knot is called tame if there exists a homomorphism f taking E^3 onto itself so that $f(h(S^1))$ is polygonal (i.e., the union of a finite number of closed straight-line segments called edges, whose end points are the vertices of the knot). A knot is called wild if it is not equivalent to a polygonal knot. A knot Γ is usually specified by a projection of E^3 onto the xy-plane. Consider the parallel projection $\pi : E^3 \longrightarrow E^2 \times \{0\} \subset E^3$ defined by $\pi(x, y, z) = (x, y, 0)$. A point $p \in \pi(\Gamma)$ is called a multiple point if $\pi^{-1}(p)$ contains more than one point of Γ. The order of $p \in \pi(\Gamma)$ is the cardinality of $\pi^{-1}(p) \cap \Gamma$. Thus a double point is a multiple point of order 2.

A polygonal knot Γ is in regular position if: (i) the only multiple points of Γ are double points, and there are only a finite number of them; (ii) no double point is the image of any vertex of Γ. Each double point of the projected image of a polygonal knot in regular position is the image of two points of the knot. The one with the larger z-coordinate is called an overcrossing and the other is the corresponding undercrossing. It can be easily shown that any polygonal knot Γ is equivalent under an arbitrarily small rotation of E^3 to a polygonal knot in regular position.

In what will follow we will only consider tame knots and hence may as well assume that the given knots are in regular position. Similar procedures to those to be discussed here will apply, with variations, to other simple sets; for example, to links (unions of disjoint simple closed curves) or graphs (1-dimensional complexes) in 3-space, to 2-spheres in 4-space, etc. I shall now describe an algorithm for reading from a regular projection of a knot Γ a set of generators and defining relations giving a presentation of $\pi_1(S^3\text{-}\Gamma)$.

In a regular projection of Γ, the number n of double points is finite. Over each double point, Γ has an undercrossing point and an overcrossing point: the n undercrossing points divide Γ into n arcs; let x_j denote the element of $\pi_1(S^3\text{-}\Gamma)$ represented by a loop that circles once round the jth arc in the direction of a left-handed screw and doesn't do anything unusual (we have first given Γ an orientation). It is intuitively clear that $x_1, x_1, \ldots, x_n$ generate $\pi_1(S^3\text{-}\Gamma)$, and it is even not too difficult to prove.

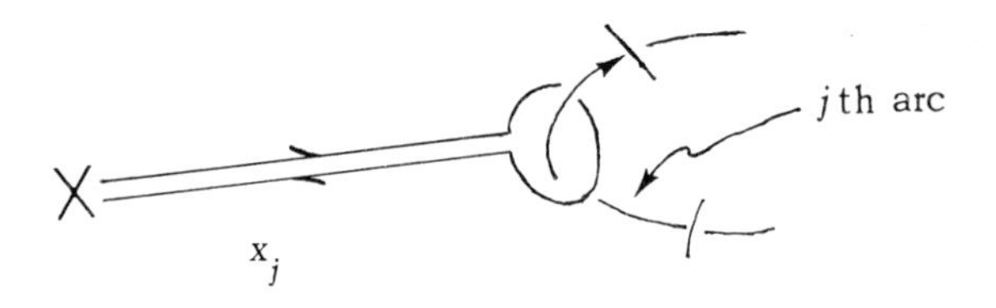

At each crossing a relation can be read.

$$x_j x_i x_j^{-1} = x_k$$

jth arc

kth arc

x_j

ith arc

x_k

x_i

The following picture show why this is a true relation.

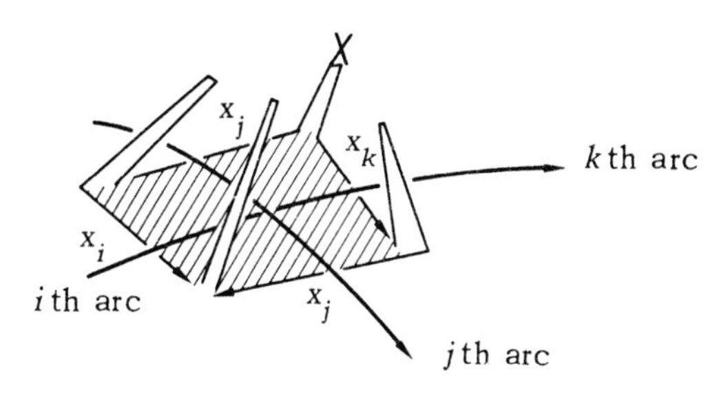

The n relations $r_1, \ldots, r_n$ obtained in this way form a complete system of defining relations; that is, any relation in $\pi_1(S^3\text{-}\Gamma)$ is a consequence of them. This fact is a little more difficult to prove (the proof only involves simple, but clever, applications of van Kampen's Theorem). We have now obtained a presentation $P = (x_1, \ldots, x_n \mid r_1, \ldots, r_n)$ of $\pi_1(S^3\text{-}\Gamma)$. That is, a symbol listing the n generators $x_1, \ldots, x_n$ and the n defining relations $r_1 = 1, \ldots, r_n = 1$. Also one can easily show that anyone of the relations $r_1 = 1, \ldots, r_n = 1$ is a consequence of the others. Thus we arrive at a presentation $(x_1, \ldots, x_n \mid r_1, \ldots, r_{n-1})$ of $\pi_1(S^3\text{-}\Gamma)$.

We now illustrate the above discussion with some simple examples.

Example 1:

a

$\pi_1(S^3\text{-}\Gamma) \cong G$ having presentation $(a \mid \;)$, hence $\pi_1(S^3\text{-}\Gamma) \cong Z$

Example 2 (trefoil knot):

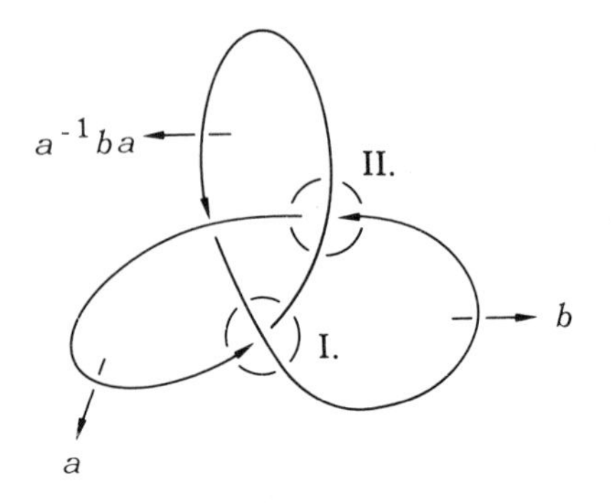

I. $ab^{-1}a^{-1}b^{-1}ab = 1$

II. $a^{-1}b^{-1}aa^{-1}a^{-1}bab = 1$

$\pi_1(S^3\text{-}\Gamma) \cong G$ having presentation $(a, b \mid aba = bab)$

Example 3 (figure eight):

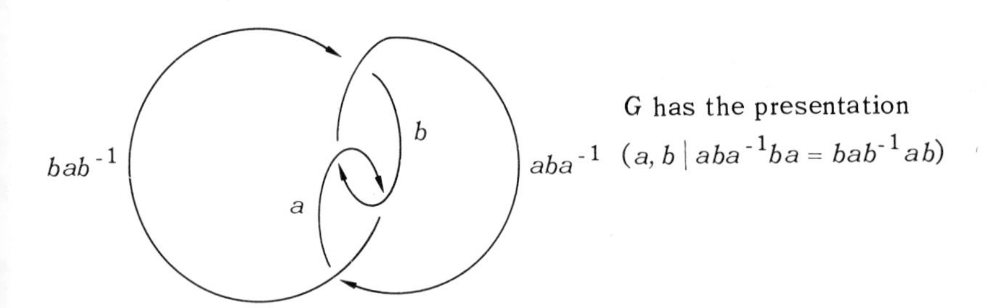

G has the presentation $(a, b \mid aba^{-1}ba = bab^{-1}ab)$

Example 4 (square knot):

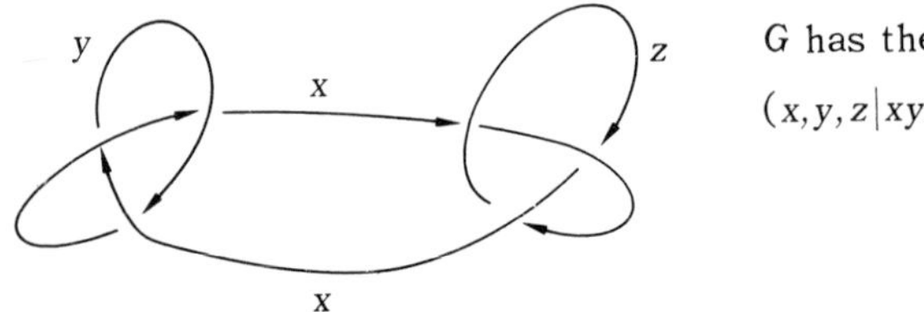

G has the presentation $(x, y, z \mid xyx = yxy, xzx = zxz)$

Example 5 (the Borromean rings):

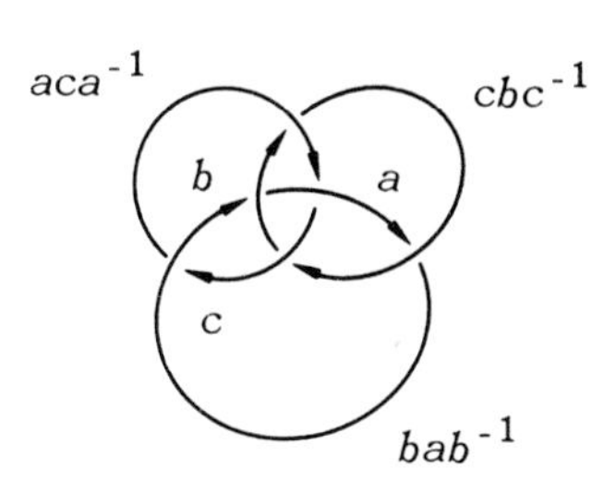

G has the presentation

$(a, b, c \mid cbc^{-1}acb^{-1}c^{-1} = bab^{-1}, \text{etc})$

Example 6:

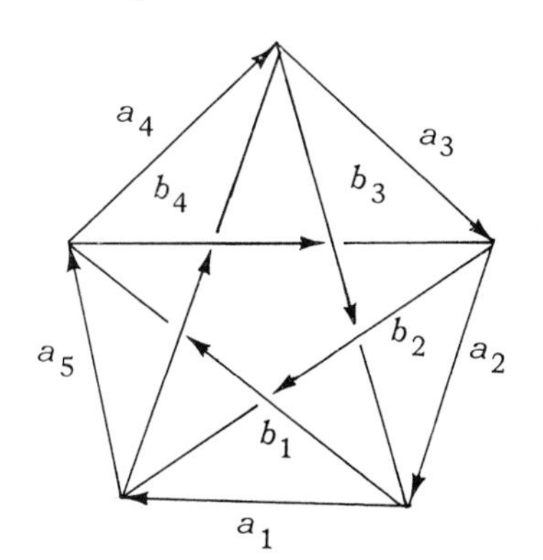

G has the presentation

$(a_1, a_2, a_3, a_4, a_5, b_1, b_2, b_3, b_4, b_5 \mid a_1 b_1 b_2 = a_2 b_2 b_3 = a_3 b_3 b_4 = a_4 b_4 b_5 = a_5 b_5 b_1)$

Remarks. In the case of knots, the abelianized group G/G' is is infinite cyclic (this can be seen from the form of the relations, or if you prefer, from the Alexander duality Theorem, making use of the fact that $G/G' \cong H_1$. The defect of G is 1; i.e., G has a finite presentation in which there is one more generator than relator. In general, G does not determine Γ).

References and Select Bibliography

1. ALEXANDER, J.W. 'On the sub-division of space by a polyhedron', *Proc. Natn. Acad. Sci. U.S.A.*, **10**, 6-8 (1924)
2. ALEXANDER, J.W. 'The combinatorial theory of complexes', *Ann. Math.* **30**, 292-320 (1930)
3. BING, R.H. 'A set is a 3-cell if its cartesian product with an arc is a 4-cell', *Proc. Am. Math. Soc.* **12**, 13-19 (1961)
4. BING, R.H. 'An alternative proof that 3-manifolds can be triangulated', *Ann. Math.* **69**, 37-65 (1959)
5. BING, R.H. 'Approximating surfaces with polyhedral ones', *Ann. Math.* **65**, 454-483 (1957)
6. BING, R.H. 'Locally tame sets are tame', *Ann. Math.* **59** 145-58 (1964)
7. BING, R.H. 'Necessary and sufficient conditions that a 3-manifold be S^3', *Ann. Math.* **68** 17-37 (1958)
8. BING, R.H. *Some aspects of the topology of 3-manifolds related to the Poincaré conjecture*, mimeographed notes, University of Wisconsin (1963)
9. BING, R.H. and KISTER, J.M. 'Taming complexes in hyperplanes', *Duke Math. J.* **31**, 491-511 (1964)
10. BROWN, M. 'A proof of the generalized Schoenflies Theorem', *Bull. Am. Math. Soc.* **66**, 74-76 (1960)
11. BROWN, M. 'Locally flat imbeddings of topological manifolds', *Ann. Math.* **75**, 331-341 (1962)
12. BROWN, M. 'The monotone union of open *n*-cells is an open *n*-cell', *Proc. Am. Math. Soc.* **12**, 812-814 (1961)
13. BROWN, M. and GLUCK, H. 'Stable structures on manifolds', *Bull. Am. Math. Soc.* **69**, 51-58 (1963)

14. CONNELL, E.H. 'Approximating homomorphisms by piecewise linear ones', **78**, 326–338 (1963)

15. CROWELL, R.H. and FOX, R.H. *Introduction to Knot Theory,* Ginn and Company, (1963)

16. CURTIS, M.L. *Combinatorial Topology* , mimeographed lecture notes taken by Peter Rice, University of Rochester (1962)

17. CURTIS, M.L. 'Cartesian products with intervals', *Proc. Am. Math. Soc.* **12**, 819-820 (1961)

18. CURTIS, M.L. 'On 2-complexes in 4-space', *Topology of 3-manifolds.* Proceedings of the 1961 Topology Institute at the University of Georgia, 204–207, Prentice-Hall (1962)

19. CURTIS, M.L. *Regular neighborhoods* , mimeographed notes, Florida State University, 1-21

20. CURTIS, M.L. and KWUN, K.W. 'Infinite sums of manifolds', *Topology* **3**, 31-42 (1965)

21. EDWARDS, C.H. Jr. 'Products of pseudo cells', *Bull. Am. Math. Soc.* **68**, 583-584 (1962)

22. EPSTEIN, D.B.A. 'Ends', *Topology of 3-manifolds,* Proceedings of the 1961 Topology Institute at the University of Georgia, 110-117, Prentice-Hall (1962)

23. FISHER, G.M. 'On the group of all homomorphisms of a manifold', *Trans. Am. Math. Soc.* **97**, 193–212 (1960)

24. FOX, R.H. 'A quick trip through knot theory', *Topology of 3-manifolds,* Proceedings of the 1961 Topology Institute at the University of Georgia, 120-167, Prentice-Hall (1962)

25. GLASER, L.C. *Contractible complexes in* S^n , Ph.D. Thesis, University of Wisconsin, 1–107 (1964)

26. GLASER, L.C. 'Contractible complexes in S^n', *Proc. Am. Math. Soc.* **16**, 1357–1364 (1965)

27. GLASER, L.C. 'Intersections of combinatorial balls of Euclidean spaces', *Bull. Am. Math. Soc.* **72**, 68–71 (1966)

28. GLASER, L.C. 'Uncountably many contractible open 4-manifolds', *Topology* **6**, 37-42 (1967)

29. GLASER, L.C. and PRICE, T.M. 'Unknotting locally flat cell pairs', *Illinois J. Math.* **10**, 425–430 (1966)

30. GLUCK, H. 'Imbeddings in the trivial range', *Bull. Am. Math. Soc.* **69**, 824-831 (1963)

31. GREATHOUSE, C.A. 'Locally flat, locally tame, and tame embeddings', *Bull. Am. Math. Soc.* **69**, 820–823 (1963)

32. GUGENHEIM, V.K. 'Piecewise linear isotopy and embedding of elements and spheres I', *Proc. Lond. Math. Soc.* **3**, 29-53 (1953)

33. HILTON, P.T. and WYLIE, S. *Homology Theory*, Cambridge University Press, Great Britain (1960)

34. HOMMA, T. ''On the imbedding of polyhedra in manifolds', *Yokohama Math.* **10**, 5-10 (1962)

35. HIRSCH, M.W. 'On the product of a contractible topological manifold and a cell', *Bull. Am. Math. Soc.* **68**, 588-589 (1962)

36. HUDSON, J.F.P. and ZEEMAN, E.C. 'On regular neighborhoods', *Proc. Lond. Math. Soc.* **14**, 719-45 (1964)

37. IRWIN, M.C. 'Combinatorial embeddings of manifolds', *Bull. Am. Math. Soc.* **68**, 25–27 (1962)

38. MAZUR, B. 'On imbeddings of spheres', *Bull. Am. Math. Soc.* **65**, 59-65 (1959)

39. MAZUR, B. 'A note on some contractible 4-manifolds', *Ann. Math.* **73**, 221-228 (1961)

40. MCMILLAN, D.R. Jr. 'Summary of results on contractible open manifolds', *Topology of 3-manifolds*, Proceedings of the 1961 Topology Institute at the University of Georgia, 100-102, Prentice-Hall

41. MCMILLAN, D.R. Jr. and ZEEMAN, E.C. 'On contractible open manifolds', *Proc. Cambridge Philos. Soc.* **58**, 221-224 (1962)

42. MILNOR, J. *On the relationship between differentiable manifolds and combinatorial manifolds,* mimeographed notes, Princeton University (1956)

43. MILNOR, J. 'Two complexes which are homomorphic but combinatorial distinct', *Ann. Math.* **74**, 575-590 (1961)

44. MILNOR, J. *Whitehead Torsion*, mimeographed notes, Princeton University (1964)

45. MOISE, E.E. 'Affine structures in 3-manifolds IV. Piecewise linear approximations on homomorphisms', *Ann. Math.* **55**, 215-222 (1952)

46. MOISE, E.E. 'Affine structures in 3-manifolds V. The triangulation theorem and Hauptvermutung', *Ann. Math.* **56**, 96-114 (1952)

47. NEWMAN, M.H.A. 'Boundaries of ULC sets in Euclidean *n*-space', *Proc. Natn. Acad. Sci. U.S.A.* **34**, 193-196 (1948)

48. NEWMAN, M.H.A. 'On the division of Euclidean *n*-space by topological (n-1)-spheres', *Proc. Soc.* **257**, 1-12 (1960)

49. NEWMAN, M.H.A. 'On the foundations of combinatory analysis situs', *Akad. Wet.*, Amsterdam, **29**, 611-641 (1926)

50. NEWMAN, M.H.A. 'On the superposition of *n*-dimensional manifolds', *J. Lond. Math. Soc.* **2**, 56-64 (1926)

51. PENROSE, R., WHITEHEAD, T.H.C. and ZEEMAN, E.C. Imbedding of manifolds in Euclidean space', *Ann. Math.* **73**, 613-623 (1961)

52. POENARU, V. 'La decomposition de l'hypercube en produit topologique', *Bull. Soc. Math. Fr.* **88**, 113-129 (1960)

53. PRICE, T.M. 'Equivalence of embeddings of *k*-complexes in Euclidean *n*-space for $n \leqslant 2k+1$', Abstract 614-121, *A.M.S. Notices* **11**, 564 (1964)

54. SMALE, S. 'Generalized Poincaré conjecture in dimensions greater than four', *Ann. Math.* **74**, 391-406 (1961)

55. SMALE, S. 'On the structure of Manifolds', *Am. J. Math.* **84**, 387-399 (1962)

56. STALLINGS, J. 'On topologically unknotted spheres', *Ann. Math.* **77**, 490-503 (1963)

57. STALLINGS, J. 'Polyhedral homotopy-spheres', *Bull. Am. Math. Soc.* **66**, 485–488 (1960)

58. STALLINGS, J. 'The piecewise-linear structure of Euclidean space', *Proc. Camb. Phil. Soc. Math. Phys. Sci.* **58**, 481–488 (1962)

59. WHITEHEAD, J.H.C. 'On C^1-complexes', *Ann. Math.* **41**, 809–824 (1940)

60. WHITEHEAD, J.H.C. 'On subdivisions of complexes', *Proc. Camb. Phil. Soc. Math. Phys. Sci.* **31**, 69–75 (1935)

61. WHITEHEAD, J.H.C. 'Simplicial spaces, nuclei and *m*-groups', *Proc. Lond. Math. Soc.* **45**, 243–327 (1939)

62. ZEEMAN, E.C. 'The generalized Poincaré conjecture', *Bull. Am. Math. Soc.* **67**, 270 (1961)

63. ZEEMAN, E.C. 'On the dunce hat', *Topology* **2**, 341–358 (1964)

64. ZEEMAN, E.C. 'A note on an example of Mazur', *Ann. Math.* **76**, 235–236 (1962)

65. ZEEMAN, E.C. *Seminar on combinatorial topology* (mimeographed notes), Institute des Hautes Etudes Scientifiques, Paris (1963)

66. ZEEMAN, E.C. 'The Poincaré conjecture for $n \geqslant 5$', *Topology of 3-manifolds* (M.K. Fort, Ed.), Prentice-Hall, New York 198–204 (1962)

67. ZEEMAN, E.C. 'Unknotting combinatorial balls', *Ann. Math.* **78**, 501–526 (1963)

INDEX

to both Volumes